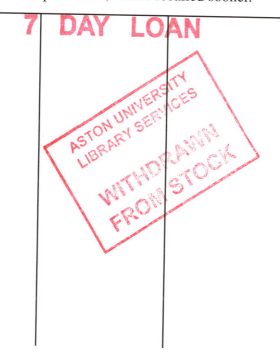

RELIABILITY ASSESSMENT OF LARGE ELECTRIC POWER SYSTEMS

**THE KLUWER INTERNATIONAL SERIES
IN ENGINEERING AND COMPUTER SCIENCE**

POWER ELECTRONICS AND POWER SYSTEMS

Consulting Editor

Tom Lipo

Other books in the series:

SPOT PRICING OF ELECTRICITY
F.C. Schweppe, M.C. Caraminis, R.D. Tabors and R.E. Bohn
ISBN 0-89838-260-2

MODERN POWER SYSTEM OPERATION AND CONTROL
A.S. Debs   ISBN 0-89838-265-3

RELIABILITY ASSESSMENT OF LARGE ELECTRIC POWER SYSTEMS

by

Roy Billinton
University of Saskatchewan

and

Ronald N. Allan
University of Manchester
Institute of Science and Technology

KLUWER ACADEMIC PUBLISHERS
Boston/Dordrecht/Lancaster

Distributors for North America:
Kluwer Academic Publishers
101 Philip Drive
Assinippi Park
Norwell, Massachusetts 02061, USA

Distributors for the UK and Ireland:
Kluwer Academic Publishers
MTP Press Limited
Falcon House, Queen Square
Lancaster LA1 1RN, UNITED KINGDOM

Distributors for all other countries:
Kluwer Academic Publishers Group
Distribution Centre
Post Office Box 322
3300 AH Dordrecht, THE NETHERLANDS

Library of Congress Cataloging-in-Publication Data

Billinton, Roy.
  Reliability assessment of large electric power
systems.
  (The Kluwer international series in engineering
and computer science. Power electronics and power
systems)
  Bibliography: p.
  Includes index.
  1. Electric power systems—Reliability.   I. Allan,
Ronald N. (Ronald Norman)   II. Title.   III. Series.
TK1005.B572   1987   621.319′1   87-37849
ISBN 0-89838-266-1

CONTENTS

PREFACE

We are very pleased to be asked to co-author this book for a variety of reasons, one of which was that it gave us further opportunity to work together. The scope proposed was very wide with the only significant proviso being that the book should be in a mongraph- style and not a teaching text. This requirement has given us the opportunity to compile a wide range of relevant material relating to present-day knowledge and application in power system reliability.

As many readers will be aware, we have collaborated in many ways over a relatively long period and have co-authored two other books on reliability evaluation. Both of these previous books were structured as teaching texts. This present book is not a discourse on "how to do reliability evaluation" but a discussion on "why it should be done and what can be done and achieved" and as such does not replace or conflict with the previous books. The three books are complementary and each enhances the others.

The material contained in this book is not specifically original since it is based on information which we have published in other forms either jointly or as co-authors with various other people, particularly our many research students. We sincerely acknowledge the important contributions made by all these students and colleagues. There are too many to mention individually in this preface but their names appear frequently in the references at the end of each chapter.

We have taken this opportunity to compile concepts and material that have appeared and have been developed in very recent years. We have been very selective and have only included material that either is now being used or can be used for practical systems. We have not included material that we consider to be still in the research domain or is not likely to have great practical significance. A particular effort has been made to include material that relates to the increased application of reliability evaluation to composite systems and the relationship between reliability and economics. In addition, we have used the IEEE Reliability Test System (RTS) for exhaustive studies. This has two distinct advantages. Firstly it will increase the understanding of the RTS so that other studies have a basis of comparison. Secondly it serves as a consistent vehicle for assessing the different techniques and application areas. The only exception to this is in the case of distribution systems since the RTS does not have a distribution system associated with it.

We are indebted to all the typists, including Jacquie Matthies, Pat Niedermuehlbichler and Sandra Calver who have contributed to the production of this book and in particular to Irene Burns for her patience and skill in producing the final copy. We would also like to acknowledge Jeffrey Billinton who did all the diagrams. We offer them all our sincerest thanks.

Finally, we would like to thank our respective wives, Joyce and Diane, for their help, encouragement and tolerance regarding the demands associated with our active research programs and the subsequent publication of the research findings.

Roy Billinton
Ron Allan

RELIABILITY ASSESSMENT OF LARGE
ELECTRIC POWER SYSTEMS

CHAPTER 1

BASIC POWER SYSTEM RELIABILITY CONCEPTS

PROBABILISTIC EVALUATION OF POWER SYSTEMS

The basic function of a power system is to supply customers, both large and small, with electrical energy as economically as possible and with an acceptable degree of reliability and quality. While satisfying this function, the power system should also remain within a set of operational constraints. Some of these constraints relate directly to the quality of supply such as busbar voltage violations and frequency variations. Other constraints, not directly seen by consumers, are equally important in an operating sense including equipment ratings, system stability limits and fault levels.

Modern society, because of its pattern of social and working habits, has come to expect that the supply should be continuously available on demand. This is not possible due to random system failures which are generally outside the control of power system engineers. The probability of customers being disconnected, however, can be reduced by increased investment during either the planning phase, operating phase or both. It is evident therefore that the economic and reliability constraints can conflict, and this can lead to difficult managerial decisions at both the planning and operating phases. The level of investment also impacts on the system operational constraints and therefore directly affects the conditions under which the system will function adequately and securely.

These problems have always been widely recognized by
power system managers, designers, planners and operators.
Design, planning and operating criteria and techniques
have been developed over many decades in an attempt to
resolve and satisfy the dilemma between the economic, reliability and operational constraints. The criteria and
techniques first used in practical applications were all
deterministically based and many of these criteria and
techniques are still in use today. The essential weakness of deterministic criteria is that they do not respond to nor do they reflect the probabilistic or stochastic nature of system behavior, of customer demands or
of component failures.

The need for probabilistic evaluation of system behavior has been recognized since at least the 1930s, and
it may be questioned why such methods have not been widely used in the past. The main reasons were lack of data,
limitations of computational resources, lack of realistic
techniques, aversion to the use of probabilistic techniques and a misunderstanding of the significance and
meaning of probabilistic criteria and indices. None of
these reasons need be valid today as most utilities have
relevant reliability databases, computing facilities are
greatly enhanced, evaluation techniques are highly developed and most engineers have a working understanding of
probabilistic techniques. Consequently, there is now no
need to artificially constrain the inherent probabilistic
or stochastic nature of a power system into a deterministic framework.

A wide range of probabilistic techniques have been
developed. These include techniques for reliability
evaluation [1-5,17], probabilistic load flow [6-8] and
probabilistic transient stability [9-11]. The fundamental and common concept behind each of these developments is the need to recognize that power systems behave
stochastically and all input and output state and event

parameters are probabilistic variables. This book however is concerned only with reliability evaluation.

Present-day studies suggest that the "worst case"
system conditions which occur very infrequently should
not be utilized as design limits or criteria because of
economic restrictions. Probabilistic techniques have
been developed which recognize, not only the severity of
a state or an event and its impact on system behavior and
operation, but also the likelihood or probability of its
occurrence. Deterministic techniques can not respond to
the latter aspect and therefore can not account objectively for stochastic variables of the system.

ADEQUACY AND SECURITY

It should be noted that the term reliability has a
very wide range of meaning and cannot be associated with
a single specific definition such as that often used in
the mission-oriented sense. It is therefore necessary to
recognize its extreme generality and to use it to indicate, in a general rather than specific sense, the overall ability of the system to perform its function. For
this reason, power system reliability assessment, both
deterministic and probabilistic, is divided into two basic aspects: system adequacy and system security as depicted in Figure 1.1. These two terms can be described
as follows.

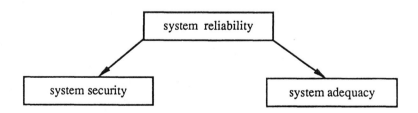

Figure 1.1 - Subdivision of system reliability

Adequacy relates to the existence of sufficient facilities within the system to satisfy the consumer load
demand or system operational constraints. These include
the facilities necessary to generate sufficient energy
and the associated transmission and distribution facilities required to transport the energy to the actual consumer load points. Adequacy is therefore associated with
static conditions which do not include system disturbances.

Security relates to the ability of the system to respond to disturbances arising within that system. Security is therefore associated with the response of the
system to whatever perturbations it is subject. These
include the conditions associated with both local and
widespread disturbances and the loss of major generation
and transmission facilities.

It is important to realize that most of the probabilistic techniques presently available for reliability
evaluation are in the domain of adequacy assessment.
Probabilistic load flow also generally falls into the adequacy domain. The ability to assess security is therefore very limited. Probabilistic transient stability
lies in this domain together with those techniques [2]
for quantifying unit commitment and response risks. The
reason for this limitation is due to the complexities associated with modeling the system in the security domain.
Consequently most of the evaluated indices are adequacy
indices and not overall reliability indices. These indices are therefore not directly comparable with those that
are collected as part of the fault reporting scheme of a
real system. In this case the indices encompass the effect of all system faults and failures irrespective of
cause and therefore include the effects of insecurity as
well as those due to inadequacy. This distinction is an
important one to recognize.

NEED FOR POWER SYSTEM RELIABILITY EVALUATION

The economic, social and political climate in which the electric power supply industry now operates has changed considerably during the last few decades. In the period between 1945 and the end of the 1950s, planning for the construction of generating plant and facilities was basically straightforward, plant construction was relatively uncomplicated, lead times were relatively small, and costs were relatively stable. This situation changed in the mid-1970s. Inflation and the huge increase in oil prices created a rapid increase in consumer tariffs and fluctuating growth patterns. Their combined effects introduced considerable uncertainty in predicting future demand.

It became evident in the late 1970s that nuclear power was not going to be the universal panacea for our electric energy needs. Environmental considerations created by public concern have been added to construction difficulties and safety and reliability problems. Conservation also became a major issue and renewable energy sources such as wind, solar, wave, etc. have and are being considered for replacing some of the energy generated by fossil-fired stations.

The only way in which all these competing and diverse uncertainties can be weighted together in an objective and consistent fashion, is by the use of quantitative reliability evaluation techniques. The results can then be related to the economic aspects of system planning and operation, the impact of which is playing an increasing role in present and future power system developments.

One very important outcome of the economic situation in which the electric power supply industry finds itself is that it is being scrutinized very closely by many different organizations and individuals. The industry is capital intensive; it plays a major role in the economic and social well-being of a nation and indeed on the qual-

ity of life. Governments, licensing bodies, consumer ad-
vocates, environmental and conservation groups and even
private citizens are expressing their concerns in ways
which did not exist a decade ago. It is with this background that the present reliability techniques and concepts are being developed, utilized and scrutinized.

It is therefore very useful to give some thought to
defining the problem zones in the general areas of overall power system reliability evaluation and to discuss
data needs, methodologies and techniques. It is necessary, however, to first create a common philosophy on a
broad scale which can also be articulated to those outside the industry as well as those within it.

FUNCTIONAL ZONES

The basic techniques for adequacy assessment can be
categorized in terms of their application to segments of
a complete power system. These segments are shown in
Figure 1.2 and are defined as functional zones: generation, transmission and distribution. This division is
the most appropriate as most utilities are either divided
into these zones for purposes of organization, planning,
operation and/or analysis or are solely responsible for

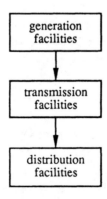

Figure 1.2 - Basic functional zones

one of these functions. Adequacy studies can be, and
are, conducted individually in these three functional
zones.

HIERARCHICAL LEVELS

The functional zones shown in Figure 1.2 can be combined to give the hierarchical levels shown in Figure
1.3. These hierarchical levels can also be used in adequacy assessment. Hierarchical Level I (HLI) is concerned only with the generation facilities. Hierarchical
Level II (HLII) includes both generation and transmission
facilities and HLIII includes all three functional zones
in an assessment of consumer load point adequacy. HLIII
studies are not usually done directly due to the enormity
of the problem in a practical system. Instead, the
analysis is usually performed only in the distribution
functional zone in which the input points may or may not
be considered fully reliable. Functional zone studies
are often done which do not include the functional zones
above them. These are usually performed on a subset of

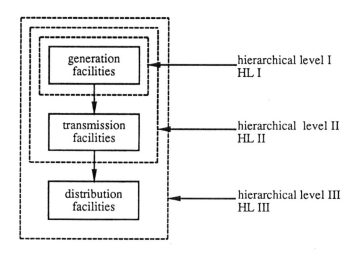

Figure 1.3 - Hierarchical levels

the system to examine a particular configuration or topological change. These analyses are frequently undertaken
in the subtransmission and distribution system functional
zones because these are less affected by the geographical
location of the generating facilities.

ADEQUACY EVALUATION

HLI Studies

At HLI the total system generation is examined to determine its adequacy to meet the total system load requirement. This is usually termed "generating capacity
reliability evaluation". The system model at this level
is shown in Figure 1.4.

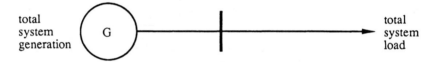

Figure 1.4 - Model for hierarchical level I

In HLI studies, the transmission system and its ability to move the generated energy to the consumer load
points is ignored. The only concern is in estimating the
necessary generating capacity to satisfy the system demand and to have sufficient capacity to perform corrective and preventive maintenance on the generating facilities. The historical technique used for determining
this capacity was the percentage reserve method. In this
approach the required reserve is set as a fixed percentage of either the installed capacity or the predicted
load. Other criteria, such as a reserve equal to one or
more largest units, have also been used. These deterministic criteria have now been largely replaced by probabilistic methods which respond to and reflect the actual
factors that influence the reliability of the system.
Criteria such as loss of load expectation (LOLE), loss of

energy expectation (LOEE), and frequency and duration (F&D) can be used.

The LOLE is the average number of days on which the daily peak load is expected to exceed the available generating capacity. It therefore indicates the expected number of days on which a load loss or deficiency will occur. It can be extended to predict the number of hours in which a deficiency may occur. It does not indicate the severity of the deficiency and neither does it indicate the frequency nor the duration of loss of load. Despite these shortcomings, it is the most widely used probabilistic criterion in generation planning studies.

The LOEE is the expected energy that will not be supplied by the generating system due to those occasions when the load demanded exceeds available generating capacity. This is a more appealing index for two reasons. Firstly it measures severity of deficiencies rather than just number of occasions. Consequently the impact of energy shortfalls as well as their likelihood is evaluated. Secondly, because it is an energy based index, it reflects more closely the fact that a power system is an energy supply system. It is therefore likely that it will be used more widely in the future particularly for those cases when alternative energy replacement sources are being considered. The complementary value of energy not supplied, i.e. energy actually supplied, is sometimes divided by the total energy demanded. This gives a normalized index known as the energy index of reliability (EIR) which can be used to compare the adequacy of systems that differ considerably in size.

The F&D criterion is an extension of the LOLE index in that it also identifies the expected frequency of encountering a deficiency and the expected duration of the deficiencies. It therefore contains additional physical characteristics which makes it sensitive to further parameters of the generating system, and so it provides

more information to power system planners. The criterion
has not been used very widely in generating system reliability analyses, although it is extensively used in network studies.

These indices are generally calculated using direct
analytical techniques although Monte Carlo simulation can
and is sometimes used. Analytical techniques represent
the system by a mathematical model and evaluate the reliability indices from this model using mathematical solutions. Monte Carlo simulation methods, however, estimate
the reliability indices by simulating the actual process
and random behavior of the system. The method therefore
treats the problem as a series of real experiments. There
are merits and demerits in both methods. Monte Carlo
simulation usually requires a large amount of computing
time and is not used extensively if alternative analytical methods are available. In theory, however, it can
include any system effect or system process which may
have to be approximated in analytical methods. It can
also evaluate indices and probability distribution functions that analytical methods can not. The application
and use of both analytical techniques and Monte Carlo
simulation are described in several of the subsequent
chapters of this book.

The LOLE approach is by far the most popular and can
be used for both single systems and interconnected systems. Expectation indices are most often used to express
the adequacy of the generation configuration. These indices give a physical interpretation which cannot be provided by a probability value. There is, however, considerable confusion both inside and outside the power industry on the specific meaning of these expectation indices
and the use that can be made of them. A single expected
value is not a deterministic parameter. It is the mathematical expectation of a probability distribution (albeit unknown), i.e. it is therefore the long-run average

value. These expectation indices provide valid adequacy
indicators which reflect factors such as generating unit
size, availability, maintenance requirements, load characteristics and uncertainty, and the potential assistance
available from neighboring systems.

The basic modeling approach for an HLI study is shown
in Figure 1.5. The capacity model can take a number of
forms. It is formed in the direct analytical methods [2]
by creating a capacity outage probability table. This
table represents the capacity outage states of the generating system together with the probability of each state.
The load model can either be the daily peak load variation curve (DPLVC), which only includes the peak loads of
each day, or the load duration curve (LDC) which represents the hourly variation of the load. The risk indices
are evaluated by convolution of the capacity and load
models. Generally the DPLVC is used to evaluate LOLE
indices giving a risk expressed in number of days during
the period of study when the load will exceed available
capacity. If an LDC is used, the units will be the
number of hours. If LOEE indices are required, the LDC
must be used.

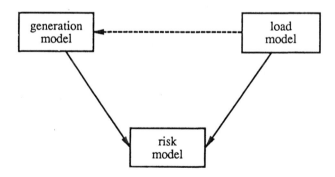

Figure 1.5 - Conceptual tasks for HLI evaluation

Limited considerations of transmission can be included in HLI studies. These include the modeling of remote generation facilities (Figure 1.6a) and intercon-

nected systems (Figure 1.6b).   In the latter case, only
the interconnections between adjacent systems are modeled; not the internal system connections.

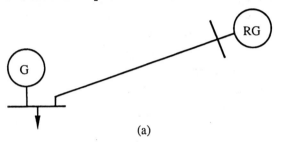

(a)

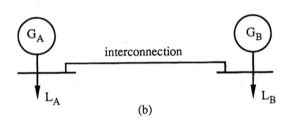

(b)

Figure 1.6 – (a) Model of remote generation in HLI stud-
ies, (b) model of interconnected systems in
HLI studies

In the case of remote generation, the capacity model
of the remote source is modified by the reliability of
the transmission link before being added to the system
capacity model.   In the case of interconnected systems
the available assistance model of the assisting system is
modified before being added to the capacity model of the
system under study.   Various aspects concerning HLI adequacy evaluation are discussed in detail in Chapter 2.

HLII Studies

In HLII studies, the simple generation-load model
shown in Figure 1.4 is extended to include bulk transmission.   Adequacy analysis at this level is usually termed

composite system or bulk transmission system evaluation.
A small composite system is shown in Figure 1.7.

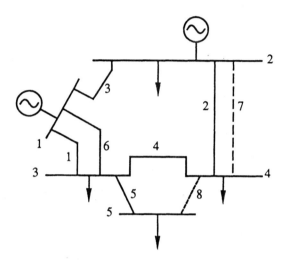

Figure 1.7 - Composite system for HLII studies

HLII studies can be used to assess the adequacy of an
existing or proposed system including the impact of various reinforcement alternatives at both the generation and
transmission levels. In the case of the system shown in
Figure 1.7, it may be necessary to evaluate the effects
of such additions as lines 7 and 8. These effects can be
assessed by evaluating two sets of indices: individual
bus (load-point) indices and overall system indices.
These indices are complementary, not alternatives. The
system indices give an assessment of overall adequacy and
the load-point indices monitor the effect on individual
busbars and provide input values to the next hierarchical
level.

The HLII indices can be calculated [2] for a single
load level and expressed on a period basis, such as a
year, or obtained using simplified load models for a specific period. Although these indices add realism by including bulk transmission, they are still adequacy indi-

cators. They do not include the system dynamics or the
ability of the system to respond to transient disturbances. They simply measure the ability of the system to
adequately meet its requirements in a specified set of
probabilistic states. There are many complications in
this type of analysis such as overload effects, redis-
patch of generation and the consideration of independent,
dependent, common-cause and station-associated outages.
Many of these aspects have not yet been fully resolved
and there is no universally accepted method of analysis.
The important aspect therefore is for each utility to
decide which parameters and effects are important in its
own case and then to use a consistent method of analysis.
A discussion on the various types of outages is presented
in Appendix 3. In addition, some of the aspects needing
consideration are discussed below.

The single-line/busbar representation shown in Figure
1.7 is a conventional load-flow representation and does
not indicate some of the complexities which should be included in the analysis. This can be illustrated by considering busbar 1 where there are three transmission
lines and a generating facility. This could have a configuration as shown in Figure 1.8. The protection system
of this ring substation is such that a single fault on
some components will cause a multiple-outage event. It
follows therefore that the total protection system should
be included in the analysis.

It is also seen in Figure 1.7 that two transmission
lines (1 and 6) leave busbar 1 and go to the same system
load point (busbar 3). If these are on the same tower
structure or possibly the same right of way, there is the
possibility of simultaneous outages due to a common-mode
event. All these factors, plus others, can be included
in an HLII evaluation. The primary indices are, however,
expected values and are highly dependent on the modeling
assumptions used in the computer simulation. Various as-

pects concerning HLII adequacy evaluation are discussed
in detail in Chapter 3.

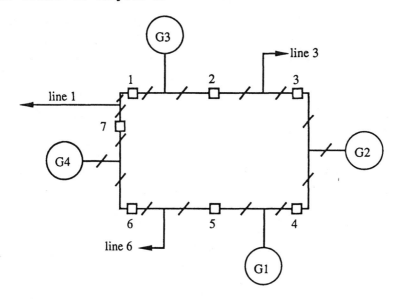

Figure 1.8 - Typical substation configuration

HLIII Studies

The overall problem of HLIII evaluation can become
very complex in most systems because this level involves
all three functional zones, starting at the generating
points and terminating at the individual consumer load
points.

For this reason, the distribution functional zone is
usually analyzed as a separate entity. The HLIII indices
can be evaluated, however, by using the HLII load-point
indices as the input values at the sources of the distribution functional zone being analyzed. The objective of
the HLIII study is to obtain suitable adequacy indices at
the actual consumer load points. The primary indices are
the expected frequency (or rate) of failure, the average
duration of failure and the annual unavailability (or
outage time) of the load points. Additional indices, such

as expected load disconnected or energy not supplied, can
also be obtained.

The analytical methods [2] for evaluating these indices are highly developed. The usual techniques are based
on the minimal-cut-set method or failure-modes analysis
in conjunction with sets of analytical equations which
can account for all realistic failure and restoration
processes. Various aspects concerning HLIII adequacy
evaluation are discussed in detail in Chapter 4.

Discussion On HL Evaluation

The adequacy indices calculated at each hierarchical
level are physically different. Those calculated at HLI
are a measure of the ability of the generating system to
satisfy the system load requirement. The indices calculated at the HLII level are a measure of the ability of
the system to satisfy the individual load requirements at
the major load points. They extend the HLI indices by
including the ability to move the generated energy
through the bulk transmission system. The individual
customer adequacy is reflected in the HLIII indices. In
most systems, the inadequacy of the individual load
points is caused mainly by the distribution system.
Therefore the HLII adequacy indices generally have a negligible effect on the individual load point indices.
Typical statistics [12] show that HLII indices contribute
less than 1% to customer unavailability. The HLI and
HLII indices are very important, however, because failures in these parts of the system affect large sections
of the system and therefore can have widespread and perhaps catastrophic consequences for both society and its
environment. Failures in the distribution system, although much more frequent, have much more localized effects.

RELIABILITY-COST/RELIABILITY-WORTH

Adequacy studies of a system are only part of the required overall assessment. As discussed earlier, the economics of alternative facilities play a major role in the decision-making process. The simplest approach which can be used to relate economics with reliability is to consider the investment cost only. In this approach, the increase in reliability due to the various alternative reinforcement or expansion schemes are evaluated together with the investment cost associated with each scheme. Dividing this cost by the increase in reliability gives the incremental cost of reliability, i.e. how much it will cost for a per unit increase in reliability. This approach is useful for comparing alternatives when it is known for certain that the reliability of a section of the power system must be increased; the lowest incremental cost of reliability being the most cost effective. This is a significant step forward compared with assessing alternatives and making major capital investment decisions using deterministic techniques.

The weakness of the approach is that it is not related to either the likely return on investment or the real benefit accruing to the consumer, utility and society. In order to make a consistent appraisal of economics and reliability, albeit only the adequacy, it is necessary to compare the adequacy cost (the investment cost needed to achieve a certain level of adequacy) with adequacy worth (the benefit derived by the utility, consumer and society). A step in this direction is achieved by setting a level of incremental cost which is believed to be acceptable to consumers. Schemes costing less than this level would be considered viable but schemes costing greater than this level would be rejected. A complete solution however requires a detailed knowledge of adequacy worth.

This type of economic appraisal is a fundamental and

important area of engineering application, and it is possible to perform this kind of evaluation at the three hierarchical levels discussed. A goal for the future should
be to extend this adequacy comparison within the same
hierarchical structure to include security, and therefore
to arrive at reliability-cost and reliability-worth evaluation. The extension of quantitative reliability analysis to the evaluation of service worth is a deceptively
simple inclusion which is fraught with potential misap-
plication. The basic concept of reliability-cost/relia-
bility-worth evaluation is relatively simple and is summarized in Figure 1.9. This same idea can also be presented by the cost/reliability curves of Figure 1.10.

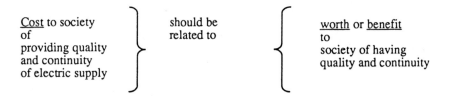

Cost to society of providing quality and continuity of electric supply } should be related to { worth or benefit to society of having quality and continuity

**Figure 1.9 - Reliability cost/reliability worth relation-
ship**

The curves in Figure 1.10 show that the utility cost
will generally increase as consumers are provided with
higher reliability. On the other hand, the consumer
costs associated with supply interruptions will decrease
as the reliability increases. The total costs to society
will therefore be the sum of these two individual costs.
This total cost exhibits a minimum, and so an 'optimum'
or target level of reliability is achieved.

This concept is quite valid. Two difficulties arise
in its assessment. First the calculated indices are usually derived only from adequacy assessments at the various hierarchical levels. Secondly, there are great problems in assessing consumer perceptions of outage costs.

The disparity between the calculated indices and the monetary costs associated with supply interruptions are
shown in Figure 1.11.

The left-hand side of Figure 1.11 shows the calculated indices at the various hierarchical levels. The
right-hand side indicates the interruption cost data obtained by user studies. It is seen that the relative disparity between the calculated indices at the three hierarchical levels and the data available for worth assessment decreases as the consumer load points are ap-
proached.

There have been many studies concerning interruption
and outage costs [13]. These studies show that, although
trends are similar in virtually all cases, the costs vary
over a wide range and depend on the country of origin and
the type of consumer. It is apparent therefore that there
is still considerable research needed on the subject of
the cost of an interruption. This research should consider the direct and the indirect costs associated with
the loss of supply, both on a local and widespread basis.
Various aspects concerning reliability cost/reliability
worth are discussed in detail in Chapter 5.

RELIABILITY DATA

Any discussion of quantitative reliability evaluation
invariably leads to a discussion of the data available
and the data required to support such studies. Valid and
useful data are expensive to collect, but it should be
recognized in the long run that it will be even more expensive not to collect them. It is sometimes argued as
to which comes first: reliability data or reliability
methodology. Some utilities do not collect data because
they have not fully determined a suitable reliability
methodology. Conversely, they do not conduct reliability
studies because they do not have any data. It should be
remembered that data collection and reliability evalua-

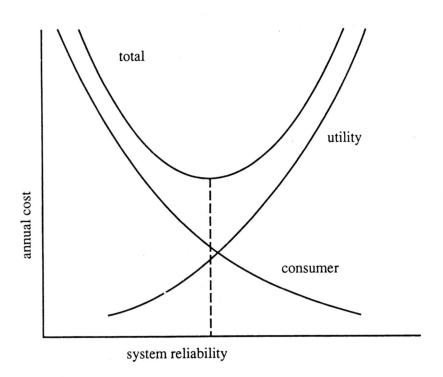

Figure 1.10 – Reliability costs

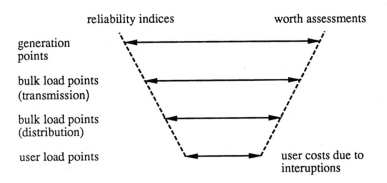

**Figure 1.11 – Disparity between indices and worth at dif-
ferent hierarchical levels**

tion must evolve together and therefore the process is
iterative. The point at which to stop on either should
be based on the economic use to be made of the tools,
techniques and data. We firmly believe, however, that
this point has not yet been reached in either sector.

When collecting data, it should be remembered that an
unlimited amount of data can be collected. It is ineffi-
cient and undesirable to collect, analyze and store more
data than is required for the purpose intended. It is
therefore essential to identify how the data will be used
before deciding what data to collect.

In conceptual terms, data can be collected for one or
both of two reasons; assessment of past performance and/
or prediction of future system performance. The past
assessment looks back at the past behavior of the system
whereas the predictive procedure looks forward at future
system behavior. In order to predict however, it is essential to transform past experience into the required
future prediction. Collection of data is therefore essential as it forms the input to relevant reliability
models, techniques and equations.

It should also be remembered that the data requirements should reflect the needs of the predictive methodology. This means that the data must be sufficiently
comprehensive to ensure that the methods can be applied
but restrictive enough to ensure that unnecessary data is
not collected nor irrelevant statistics evaluated. The
data should therefore reflect and respond to the factors
that affect system reliability and enable it to be modeled and analyzed. This means that it should relate to
the two main processes involved in component behavior,
namely the failure process and the restoration process.
It cannot be stressed too strongly that, in deciding data
to be collected, a utility must make its decision on the
basis of the factors that have an impact on its own planning and design considerations.

The quality of the data and evaluated indices depends
on two important factors; confidence and relevance. The
quality of the data and thus the confidence that can be
placed in it, is clearly dependent on the accuracy and
completeness of the information compiled by operation and
maintenance personnel. It is therefore essential that
they should be made fully aware of the future use to
which the data will be put and the importance it will
play in the later developments of the system. The quality of the statistical indices is also dependent on how
the data is processed, how much pooling is done and the
age of the data currently stored. These factors affect
the relevance of the indices for future use.

RELIABILITY TEST SYSTEMS

Although reliability is recognized as one of the important parameters which must be taken into account during the design and planning phases of a power system, one
particularly frustrating aspect associated with the wide
range of material published on this subject is that, until 1979, there was no general agreement of either the
system or the data that should be used to demonstrate or
test proposed techniques. Consequently it was not easy,
and often impossible, to compare and/or substantiate the
results obtained from various proposed methods.

This problem was recognized by the IEEE Subcommittee
on the Application of Probability Methods (APM) which, in
1979, published the IEEE Reliability Test System [14],
referred to as the RTS throughout this book. This is a
reasonably comprehensive system containing generation data, transmission data and load data. It is intended to
provide a consistent and generally acceptable set of data
that can be used both in generation capacity (HLI) and in
composite system reliability (HLII) evaluation. This will
enable results obtained by different people using different methods to be compared. Sufficient data for the ap-

plications described in the various chapters of this book have been extracted from Reference 14 and this is given in Appendix 1. Various additional data has been proposed [15,16] in order to enhance the applicability of the RTS. This data is presented in Appendix 2.

The RTS is used extensively in the applications given in subsequent chapters. The only exception is in Chapter 4 which concerns distribution systems (HLIII) since the RTS does not have any distribution defined for it. This use of the RTS not only provides a consistent vehicle for describing the various applications, it also enables a comprehensive understanding of the system to be derived and presented.

CONCLUSIONS

This chapter has discussed and described various philosophical aspects concerning power system reliability and, in particular, adequacy. The framework described is one on which the discussions within the power industry and with external groups can be ideally based. The need for such a framework is already evident due to the grow-ing number of people and organizations wishing to affect the planning decisions of power systems, and this trend will expand in the future. The framework described in this chapter is also suitable for assisting in the deci-sion making process itself. The main reasons for this conclusion are as follows.

There should be some conformity between the reliabil-ity of various segments of the system and a balance is required between generation, transmission and distribu-tion. This does not imply that the reliability of each should be equal. Differing levels of reliability are justified, for example, because of the importance of a particular load, or because generation and transmission failures can cause widespread outages whereas distribu-tion failures tend to be highly localized.

There should also be some benefit gained from any improvement in reliability.   The most useful concept for
assessing this benefit is to equate the incremental or
marginal investment cost to the incremental or marginal
consumers' evaluation of the improved reliability.   The
main difficulty with such a concept is the present uncertainty in the consumers' valuation.   In the absence of
comprehensive and complete data it is beneficial for an
individual utility to arrive at some consistent criterion
using experience or judgement by which they can then assess the benefit of expansion and reinforcement schemes.

The hierarchical framework presented in this chapter
is extended in the subsequent chapters of this book.
Chapter 2 illustrates the application of basic reliabil-
ity/adequacy concepts at HLI including conventional and
non-conventional energy sources.   Chapter 3 presents the
application of basic concepts to HLII and Chapter 4 deals
with applications in the distribution functional zone.
The determination of customer interruption costs for residential, commercial and industrial consumers is presented in Chapter 5 together with the development of customer
damage functions suitable for reliability worth analysis
at HLI, II and III.

It should be noted that these chapters are not intended to be exhaustive texts on any of these topics.
Instead they are primarily concerned with a discussion
relating to latest developments and applications and to
illustrate what can be achieved in practise at the different hierarchical levels.   The reader should refer to
References 1 and 2 in order to become acquainted with the
basic details of reliability evaluation techniques.

REFERENCES

1. Billinton, R. and Allan, R.N., "Reliability Evaluation Of Engineering Systems: Concepts And Techniques," Longman, London (England)/Plenum Press, New
   York, 1983.
2. Billinton, R. and Allan, R.N., "Reliability Evalua-

tion Of Power Systems," Longman, London (England)/ Plenum Press, 1984.

3. Billinton, R., "Bibliography On The Application Of Probability Methods In Power System Reliability Evaluation," IEEE Trans., PAS-91, 1972, pp. 649-660.

4. IEEE Committee Report, "Bibliography On The Application Of Probability Methods In Power System Reliability Evaluation, 1971-1977," ibid., PAS-97, 1978, pp. 2235-2242.

5. Allan, R.N., Billinton, R. and Lee, S.H., "Bibliography On The Application Of Probability Methods In Power System Reliability Evaluation, 1977-1982," ibid., PAS-103, 1984, pp. 275-282.

6. Allan, R.N., Borkowska, B. and Grigg, C.H., "Probabilistic Analysis Of Power Flows," Proc. IEE, 121, 1974, pp. 1551-1556.

7. Allan, R.N., Leite da Silva, A.M. and Burchett, R.C., "Evaluation Methods And Accuracy In Probabilistic Load Flow Solutions," IEEE Trans. on Power Apparatus and Systems, PAS-100, 1981, pp. 2539-2546.

8. Leite da Silva, A.M., Allan, R.N. and Ariente, V.L., "Probabilistic Load Flow Considering Dependence Between Input Nodal Powers," IEEE Trans. on Power Apparatus and Systems, PAS-103, 1984, pp. 1524-1530.

9. Billinton, R. and Kuruganty, P.R.S., "A Probabilistic Index For Transient Stability," IEEE Transactions, PAS-99, 1980, pp. 195-207.

10. Kuruganty, P.R.S. and Billinton, R., "A Probabilistic Assessment Of Transient Stability," The Journal of Electric Power and Energy Systems, Volume 2, No. 2, 1980, pp. 115-119.

11. Billinton, R. and Chu, K., "Transient Stability Evaluation In An Undergraduate Curriculum - A Probabilistic Approach," IEEE Transactions, PWRS-1, No. 4, 1986, pp. 171-178.

12. Dixon, G.F.L. and Hammersley, H., "Reliability And Its Cost In Distribution Systems," IEE Conference Publication 148, 1977, pp. 81-84.

13. Billinton, R., Wacker, G. and Wojczynski, E., "Comprehensive Bibliography Of Electrical Service Interruption Costs," IEEE Transactions, PAS-102, 1983, pp. 1831-1837.

14. IEEE Committee Report, "IEEE Reliability Test System," IEEE Trans. PAS-98, 1979, pp. 2047-2054.

15. Allan, R.N., Billinton, R. and Abdel-Gawad, N.M., "The IEEE Reliability Test System - Extensions To And Evaluation Of The Generating System," IEEE Trans on Power Systems, PWSR-1, No. 4, 1986, pp. 1-7.

16. Billinton, R., Vohra, P.K and Kumar, S., "Effect Of Station Originated Outages In A Composite System Adequacy Evaluation Of The IEEE Reliability Test System," IEEE Transactions PAS-104, No. 10, October 1985, pp. 2649-2656.

17. Allan, R.N., Billinton, R., Shahidehpour, S.M. and Singh, C., "Bibliography On The Application Of Prob-

ability Methods In Power System Reliability Evaluation, 1982-87," IEEE Winter Power Meeting, New York,
February 1988.

CHAPTER 2

GENERATION SYSTEMS ADEQUACY EVALUATION

INTRODUCTION

As described in Chapter 1, the generating system can be defined as hierarchical level I (HLI) since it forms the first or basic step in the overall planning of a power system. In order to evaluate the adequacy of HLI, it is necessary to assess the ability of the generating capacity to satisfy the system load with acceptable risk. Techniques for making this assessment were initiated in the 1930's and have been continuously developed ever since. A fairly exhaustive set of relevant papers are listed in the bibliographies of References 1-4,32 and the important evaluation techniques are documented in Reference 5. It is not the purpose of this chapter to reiterate these techniques. Instead it is intended to address several related aspects which have either been developed more recently, or which have not been treated adequately in this form before or in which the stress has changed due to more recent thinking and application. These aspects include analysis of the IEEE Reliability Test System (RTS), energy-based reliability indices, novel forms of generation and Monte Carlo simulation techniques.

ANALYSIS OF THE IEEE RELIABILITY TEST SYSTEM

Background

As discussed in Chapter 1, the RTS [6] was developed in order to create a consistent and generally acceptable

system and data set that could be used in both generation
capacity and composite system reliability evaluation.
This section, which is based on Reference 7 outlines some
of the restrictions which currently exist in the generation data of the RTS. It extends the RTS by including
more factors and system conditions which may be included
in the reliability evaluation of generating systems.
These extensions create a wider set of consistent data.
It also includes generation reliability indices for the
base and extended RTS. These indices have been evaluated
[7] without any approximations in the evaluation process
and therefore provide a set of exact indices against
which the results from alternative and approximate methods can be compared. This makes the RTS more useful in
assessing reliability models and evaluation techniques.

Existing RTS Data

The original RTS was developed in order to provide
system data that was perceived to be sufficient at that
time. Experience has shown that certain additional data
in both the generation system and the transmission system
would be desirable. This has become evident because of
developments that have taken place subsequently in both
modeling and evaluation and also because of the type and
scope of more recently published analyses. This chapter
is concerned with expanding the information relating to
the generation system [7].

Although the system input data is already comprehensive, several important aspects are omitted. These include unit derated (partial output) states, load forecast
uncertainty, unit scheduled maintenance and the effect of
interconnections. It is desirable for these factors to
be specified for the RTS in order that all users of the
RTS may consider the same data and therefore evaluate results which can easily be compared. This additional data
is specified in Appendix 2. There are many other aspects

which could also be considered including start-up failures and outage postponability. Inclusion of these, however, requires additional modeling assumptions which are
outside the scope of the present RTS and therefore,
should be included only when the RTS is revised.

System Reliability Indices

Although the original RTS paper [6] included the input data, no information was included concerning the system reliability indices. Experience has shown that this
was an important omission which should be rectified. The
main reason is that most practical evaluation techniques
include approximations and modeling assumptions regarding
the generating capacity model, the load model and/or the
evaluation algorithm.

Consider first the generating capacity model. This
has 1872 states if no rounding is used and the model is
truncated at a cumulative probability of 1×10^{-8}. In
practice, rounding and higher truncation values of cumulative probability are frequently used. Also, other approximate methods, such as the cumulant method [8] have
been developed. All of these aspects introduce approximations.

Consider now the load model. The specified RTS load
model has 364 levels if only the daily peak loads are
used and 8736 levels if the hourly loads are used. In
many practical applications, the load model is usually
represented, not by the actual daily or hourly levels,
but by a smoothed curve depicted by a restricted number
of coordinate points. These practical models are known
[5] as the daily peak load variation curve (DPLVC) in the
case of daily peak loads and the load duration curve
(LDC) in the case of hourly loads. These two curves
(DPLVC and LDC) for the RTS are shown in Figure 2.1.
This modeling aspect also leads to approximations.

It is evident from the above reasoning that a result

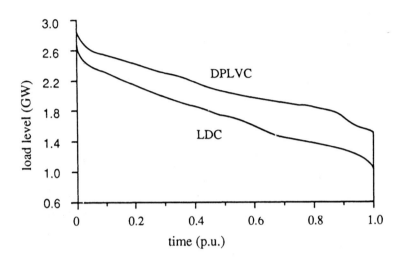

Figure 2.1 - Basic RTS load models

obtained from a particular analysis, in which one or more
of the above approximations are incorporated, will not be
exact. The degree of error however is unknown unless the
result can be compared against an exact value. Therefore
it was decided [7] that a series of results should be
evaluated for the RTS in which no approximations in the
evaluation process and no assumptions, additional to
those already associated with the RTS, were made. These
indices could then form base values against which results
from alternative and approximate methods can be compared.
All the results in the following sections, except those
showing the effect of rounding, are therefore exact (for
the given data) since no approximations have been made in
either the capacity model or the load model. They are
therefore reproducible within the precision limitations
of a particular computer. The methods used to ensure the
exactness are described in the relevant section that follows.

LOLE Analysis Of Base Case

The base case is considered to be the system as pub-

lished in the original RTS [6]. In order to evaluate the
exact loss of load expectation (LOLE) indices, the complete capacity model was developed with no rounding and
truncated at a cumulative probability of 1×10^{-8}, i.e. it
consisted of 1872 states. It was assumed that there were
no energy or capacity limitations associated with the
hydro units. The load model was represented by all 364
daily peak loads in order to evaluate the exact LOLE in
day/yr and by all 8736 hourly peak loads in order to
evaluate the exact LOLE in hr/yr. The LOLE indices were
evaluated by deducing the risk for each of these load
levels and summing over all load levels, i.e.:

$$LOLE = \sum_{i=1}^{n} P(C < L_i) \qquad (2.1)$$

where $P(C < L_i)$ = probability of loss of load on day i or
                    during hour i. This value is given di-
                    rectly by the capacity model,
and   n = 364 or 8736 as appropriate.
Using this technique, the exact LOLE indices are:
    LOLE = 1.36886 day/yr using daily peak loads,
    LOLE = 9.39418 hr/yr using hourly loads.
These two values must be considered as two fundamental
indices. They can be used to compare results obtained
using the DPLVC and the LDC respectively. Any deviation
from these values will be due to precision (or lack of
it) in the computer system.

Effect Of Rounding

If the generation model and/or load model is rounded,
the values of LOLE will differ from those given in the
previous section. This effect is illustrated in this
section in which both the generation model and the load
model have been rounded separately and concurrently.
    The generation model was rounded in steps between 20

and 100 MW using conventional techniques [5]. In practice, rounding is normally done during the development of
the model. However the results are then dependent on the
order of adding units and how frequently rounding is performed. In order to ensure a set of consistent and reproducible results, the table was rounded after the complete capacity model was evaluated. The results then become independent of the order of adding units to the
model.

The load model was rounded by first dividing the load
into n equally spaced load levels. The number of days
that each of these load levels is exceeded was deduced.
This process created n coordinate points for the DPLVC,
which was used with the capacity model to give the value
of LOLE in day/yr using conventional techniques [5]. If
a capacity level existed between two sets of coordinate
points, interpolation was used to find the number of corresponding days.

The results are shown in Table 2.1 for the cases of
rounded generation/exact load, exact generation/rounded
load and rounded generation/rounded load. It is seen
that, as severity of rounding is increased, the values of
LOLE tend to increase. It should be noted that the 364
point rounded load model is represented by 364 equally
spaced load levels which is not the same as the exact 364
individual daily peaks.

Effect Of Derated States

Derated or partial output states can have a significant effect on the LOLE, particularly units of large
capacity. There are various ways in which such units can
be included in the analysis; the only exact way being to
represent the unit by all its states and to add the unit
into the capacity model as a multi-state unit. An EFOR
(equivalent forced outage rate) representation is not an
equivalent and gives pessimistic values of LOLE [5]. It

Table 2.1 - Effect of rounding

Capacity model rounding interval (MW)	Load model (no. of points)	LOLE (d/yr)
20	exact	1.38587
40	exact	1.37978
60	exact	1.39806
80	exact	1.37687
100	exact	1.41622
exact	10	1.74649
exact	100	1.42843
exact	200	1.38993
exact	364	1.37256
20	100	1.43919
20	200	1.39869
20	364	1.38967
40	100	1.45041
40	200	1.41514
40	364	1.39415

is not normally necessary however to include more than one or possibly two derated states [5] to obtain a reasonably exact value of LOLE.

For these reasons, the 400 MW and 350 MW units of the RTS have been given a 50% derated state. The number of service hours (SH), derated hours (DH) and forced outage hours (FOH) are shown in Appendix 2 and were chosen so that the EFOR [9] of the units are identical to the FOR specified in the original RTS [6].

(a) 400 MW unit: SH = 1100 hr, DH = 100 hr   FOH = 100 hr
derated capacity = 200 MW   EH = $\dfrac{200}{400}$ DH

= 50 hr

$$\therefore \; EFOR = \frac{FDH + EH}{SH + DH + EH} = \frac{100 + 50}{1100 + 100 + 50}$$

$$= 0.12$$

(b) 350 MW unit: SH = 1150 hr   DH = 60 hr   FOH = 70 hr
derated capacity = 175 MW   EH = $\dfrac{175}{350}$ DH

= 30 hr

and EFOR = $\dfrac{70 + 30}{1150 + 70 + 30}$ = 0.08

These values of state hours give the limiting state probabilities [10] shown in Table 2.2.

Table 2.2 - Limiting state probabilities

		State probability	
Unit	Up	Derated	Down
400 MW	0.846154	0.076923	0.076923
350 MW	0.898438	0.046875	0.054687

The LOLE was evaluated using the exact generating capacity and load models i.e. no rounding, for three cases;
when one 400 MW unit, when both 400 MW units and when
both the 400 units and the 350 MW unit was represented by
3 states. The results are shown in Table 2.3.

Table 2.3 - Effect of derated states

Units derated	LOLE d/yr
1 x 400 MW	1.16124
2 x 400 MW	0.96986
2 x 400 + 1 x 350 MW	0.88258

These results show that the value of LOLE decreases
significantly when derated states are modeled. This
clearly demonstrates the inaccuracies that can be created
if EFOR values are used, particularly for the larger
units.

Effect Of Load Forecast Uncertainty

Load forecast uncertainty was modeled using a normal
distribution divided into seven discrete intervals [5].
The probabilities associated with each interval can be
evaluated [10] as the area under the density function and
these are shown in Appendix 2. It is suggested that a

load forecast uncertainty having a standard deviation of
5% should be associated with the RTS.   This is the value
specified in Appendix 2.   In the present analysis however, standard deviations from 2-15% have been considered.   The results using the exact capacity and load models are shown in Table 2.4.

Table 2.4 - Effect of load forecast uncertainty

Uncertainty %	LOLE d/yr
2	1.45110
5	1.91130
10	3.99763
15	9.50630

These results clearly show the very significant increase in LOLE as the degree of uncertainty is increased.

Effect Of Scheduled Maintenance

There are two main aspects relating to scheduled
maintenance.   The first is to ascertain or deduce the
schedule.   The second is its effect on LOLE.   The value
of LOLE will increase when maintenance is considered because of the reduced and variable reserve at different
times of the year.

The schedule selected is Plan 1 of Reference 11. This
complies with the maintenance rate and duration of the
original RTS [6], and was derived using a levelized risk
criteria.   The schedule is shown in Appendix 2.

The analysis proceeds by using the exact capacity
model and the exact load model for each week of the year.
The LOLE for each week is evaluated using Equation 2.1.
The annual LOLE is deduced by summing all the weekly values.   The details of this exercise is shown in Table 2.5
together with the overall annual LOLE. These results show
that the risk is approximately doubled when this maintenance schedule is included.

Table 2.5 - Effect of scheduled maintenance

Week nos.	LOLE d/yr
1,2,19,23-25,44-52	1.12026
3-5	0.11395
6,7	0.06801
8	0.07424
9	0.02122
10	0.04624
11	0.07223
12,13	0.04632
14	0.03701
15	0.04654
16,17	0.07203
18	0.04392
20	0.06214
21,22	0.07202
26	0.06483
27	0.02015
28	0.06718
29	0.03259
30	0.04878
31,32	0.08787
33	0.05896
34	0.02059
35	0.11809
36	0.02266
37	0.07039
38,39	0.05062
40	0.02819
41,42	0.03858
43	0.04098
Total LOLE	2.66659 d/yr

Effect Of Peak Load

One particular criticism levelled against the RTS is
that the transmission system is too reliable compared
with that of the generation system. This is because the
generation is particularly unreliable: the LOLE is
1.36886 day/yr compared with a frequently quoted practical value of 0.1 day/yr. The reason for this high level
of risk can be viewed as being due to a load level that
is too great for the generating capacity or a generating
capacity that is too small for the expected load.

The first of these reasons, i.e. the effect of the

peak load on the LOLE, is considered in this section.
Taking the RTS peak load of 2850 MW as 1 p.u., a range of
peak loads between 0.84 and 1.1 p.u. were studied.   In
each case, all 364 daily peak loads were multiplied by
the same p.u. factor and the LOLE evaluated using
Equation 2.1 and the exact capacity and load models.   The
results are plotted in Figure 2.2 and some are tabulated
in Table 2.6.   These results show that the peak load
carrying capability, PLCC [5], for a risk level of 0.1
day/yr is 2483.5 MW which is 0.8714 p.u. of the specified
[6] RTS peak load.

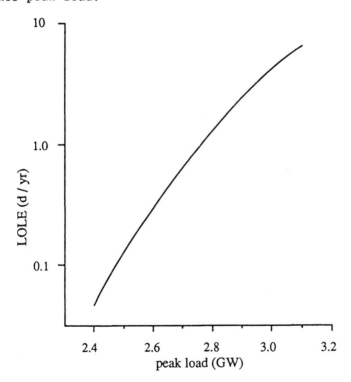

Figure 2.2 - Effect of peak load

Effect Of Additional Generation

The second reason mentioned in the previous section
for the unreliable generating system can be alleviated by

adding generating units to the system.   This was achieved
by adding a number of gas turbines each rated at 25 MW
and having a FOR of 0.12.

Table 2.6 - Effect of peak load

Multiplying factor p.u.	Peak load MW	LOLE d/yr
1.10	3135	6.68051
1.06	3021	3.77860
1.04	2964	2.67126
1.00	2850	1.36886
0.96	2736	0.65219
0.92	2622	0.29734
0.88	2508	0.12174
0.84	2394	0.04756

Using the exact capacity and load models together
with Equation 2.1 gives the results plotted in Figure 2.3
some of which are shown in Table 2.7.   These results show
that 15 such gas turbines are required in order to
achieve a PLCC of 2850 MW with a LOLE of about 0.1
day/yr.

It is therefore suggested that the generating system
of the RTS as originally specified [6] should be used as
the base but that additional 25 MW gas turbines as specified in Appendix 2 should be included if a smaller risk
index is required, e.g. in order to make transmission and
generation more comparable or to achieve an LOLE that is
nearer to frequently quoted practical values.   In order
to be consistent with the RTS and to conduct network
analysis, it is necessary to specify the busbars to which
additional generation must be attached.   This requires a
composite reliability study (see Chapter 3) but is beyond
the objective of the present chapter.

Energy Based Indices

The most popular generation reliability index is the
Loss of Load Expectation (LOLE) as derived in the previ-

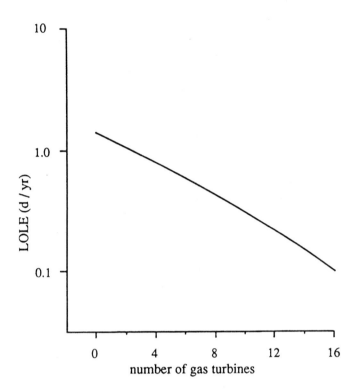

Figure 2.3 - Effect of added generation

Table 2.7 - Effect of adding gas turbines

No. of gas turbines	LOLE d/yr
1	1.18293
3	0.86372
5	0.62699
8	0.38297
10	0.27035
12	0.18709
15	0.10674
16	0.08850

ous sections. As discussed later in this chapter, energy based indices are now receiving more attention particularly for systems that have energy limitations (e.g. pumped storage) or for studying replacement of thermal

energy by novel forms of generation (e.g. solar, wind).
It is useful therefore to evaluate relevant energy indices for the RTS, these include Loss of Energy Expectation
(LOEE) and Energy Index of Reliability (EIR). An additional advantage given by energy based evaluation methods
is that the energy generated by each unit can be
evaluated [5].   This enables production costs to be
found.

The principle used to evaluate [7] exact values of
energy not supplied was as follows. Each hourly load
level is numerically equal to the energy demanded during
that hour. Consequently the total energy demanded by the
system is numerically given by the summation of all 8736
load levels. The energy not supplied can be found using
a similar principle. For each state of the capacity model
C_k, the energy not supplied E_k is given numerically by
summing all positive values of (L_i-C_k) where L_i is the
i-th load level and i=1 to 8736. The expected energy not
supplied is then given by:

$$EENS = \sum_{k=1}^{n} E_k P_k \qquad\qquad (2.2)$$

where P_k = probability of capacity state C_k,
and   n = number of capacity states.

As described in more detail later in this chapter,
this value of EENS can be evaluated after adding each
unit into the system capacity model. Hence the expected
energy produced by each unit is given [5] by the difference in EENS before and after adding the unit. The order
of adding units is important and must follow the merit
order. When all units have been added, the final value
of EENS gives the system LOEE. Also the EIR is given by:

$$EIR = 1 - LOEE/energy\ demanded \qquad\qquad (2.3)$$

Using the above principle, the expected energy pro-

duced by each unit is shown in Table 2.8.   The merit order was assumed to be that shown in the table.   Also:

energy demanded = 15297.075 GWh,

LOEE = 1.176 GWh,

EIR = 0.999923.

Table 2.8 - Energy supplied by each unit

Merit order	Type	Size MW	Energy supplied GWh
1	hydro	50	432.432
2	hydro	50	432.432
3	hydro	50	432.432
4	hydro	50	432.432
5	hydro	50	432.432
6	hydro	50	432.432
7	nuclear	400	3075.072
8	nuclear	400	3067.682
9	coal 1	350	2521.737
10	coal 2	155	963.742
11	coal 2	155	833.633
12	coal 2	155	677.731
13	coal 2	155	527.295
14	coal 3	76	217.557
15	coal 3	76	186.101
16	coal 3	76	153.884
17	coal 3	76	122.912
18	oil 1	197	196.003
19	oil 1	197	96.639
20	oil 1	197	40.645
21	oil 2	100	9.859
22	oil 2	100	5.661
23	oil 2	100	3.119
24	oil 3	12	0.268
25	oil 3	12	0.248
26	oil 3	12	0.229
27	oil 3	12	0.210
28	oil 3	12	0.194
29	GT	20	0.265
30	GT	20	0.234
31	GT	20	0.205
32	GT	20	0.181

Summary

The following results were evaluated using the exact
capacity and load models together with Equations 2.1, 2.2
and 2.3 as appropriate.   They can therefore be considered

exact and can be used to compare the results of approximate methods. The details of each "case" can be found in the preceding sections.

(a) LOLE indices:

base (as per RTS [6])	1.36886 day/yr
	[9.39418 hr/yr]
with derated states	0.88258 day/yr
with 5% load forecast uncertainty	1.91130 day/yr
with maintenance	2.66659 day/yr
with 15 x 25 MW gas turbines	0.10674 day/yr

(b) Energy indices for base system:

LOEE	1.176 GWh
EIR	0.999923

This section has extended the data and available information relating to the RTS. In so doing, it increases the value of the RTS in two important respects.

The first is that the increased range of data will enable users of the RTS to employ a consistent set of data even with extended techniques and ensure that comparison of results is much easier.

The second is that the generically exact indices for a wide range of conditions will enable the results from alternative and approximate methods to be compared since the indices quoted should be reproducible within the precision limitations of the computers used.

GENERATION PLANNING USING ENERGY BASED INDICES

Background

Although, as stated in the previous section, LOLE is still the most widely used index in the adequacy assessment of HLI, there is growing interest in the advantages of energy-based reliability indices. In fact, several utilities are already evaluating these either as primary indices or in association with indices based on LOLE techniques. This developing interest reflects the need to study energy replacement strategies and the fact that energy related indices are more responsive to changes in energy systems.

This section, which is based on References 12 and 13 describes a reliability evaluation technique for generating systems which enables energy-based indices to be derived. The great benefit of this technique is that the indices not only reflect and respond to the energy demands imposed on the system but can also be used directly in an economic assessment of the production cost of each unit and of the system. Consequently, the single technique enables two fundamental parameters of the decision-making process to be evaluated simultaneously. Alternative expansion plans can then be compared easily and directly. The technique is illustrated using the RTS.

Reliability Evaluation In Generation Planning

Several decisions have to be made in the planning phase of generation systems. One of these is to decide how much additional generating capacity in excess of the predicted load must be installed. This additional capacity, known as reserve, is needed to compensate for generating plant failures and scheduled maintenance and to compensate for load growths in excess of the prediction. If the reserve is insufficient, the reliability of supply will be very poor but if the reserve is too large, the system will be very uneconomic.

Historically, the reserve has been decided using subjective criteria such as "percentage reserve margins" and/or the "loss of one or more largest units". These criteria can not and do not take into account many of the system constraints such as generating unit failure rates and repair times, load forecast uncertainty, scheduled maintenance times of different unit sizes, etc. Consequently they are crude methods. For this reason most utilities are now using some form of probability evaluation based upon modern reliability evaluation techniques

[5,10]. Care must be taken to choose the appropriate
method since, to be meaningful and consistent, the method
must reflect and respond to the system parameters and operating criteria.

The most frequently used technique is the Loss of
Load Expectation (LOLE) method [5]. This generally uses
the Daily Peak Load Variation Curve (DPLVC) which represents the number of days that the daily peak load will be
exceeded during the period of study. Alternatively the
Load Duration Curve (LDC) can be used which represents
the number of hours that the load will be exceeded during
the period of study.

If the DPLVC is used, the evaluation index of LOLE
represents the expected number of days that a loss of
load will be encountered. If the LDC is used, the LOLE
represents the expected number of hours that a loss of
load will be encountered.

The LOLE is a very valuable technique since it enables consistent comparison to be made between alternative expansion plans. It does have two significant demerits however. Firstly, it does not indicate the severity of loss of supply and secondly it is not an energy-
based index. The latter factor is important since generating systems are energy systems and therefore it is desirable for the reliability index to reflect the energy
basis.

These problems can be overcome by using energy indices rather than load (or capacity) indices. A consequential advantage of evaluating energy indices is that they
can be used to evaluate production cost and therefore
they bring together the two primary factors of reliability and economics. For this reason, utilities are now
beginning to recognize the great merits of energy-based
reliability indices in generation planning and increased
use of these is envisaged in the near future.

Production Cost Evaluation

Production cost evaluation is an important aspect in generation planning since units of the same or similar capacity and reliability indices can have very different operating costs due to their primary fuel requirements. Historically this is usually considered as a separate exercise to the evaluation of reliability since the models and evaluation techniques are generally different. Using energy-based evaluation methods for both reliability and production cost prediction means that both parameters can be considered simultaneously which leads to an improved and more consistent decision-making process.

The most frequently used production cost method uses the LDC referred to previously. The generating units are placed in merit order and their mean or average capacity is evaluated by multiplying their rated capacity by their availability. The first unit in the merit order table is then placed in the base of the LDC and the energy it produces is calculated. Its production cost is given by multiplying its generated energy by its unit production cost. The next unit is then placed in the LDC above the first unit. This process is repeated until all units have been considered. The principle involved in this analysis is shown in Figure 2.4.

This method is an approximation and is not consistent. The fundamental error is that a unit does not generate its average capacity all the time. Instead it generates either its full capacity or nothing at all (partial/derated states may also exist). This aspect is not taken into account by the previous method and therefore a production costing technique based on such a method does not respond to nor reflect the true system behavior. The problem can be overcome however by the combined reliability/production cost method described in the next section.

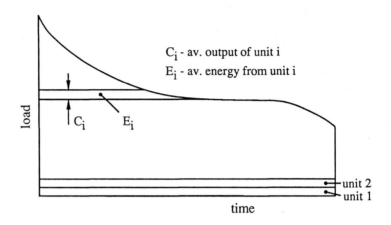

Figure 2.4 – Energy production

Energy Based Reliability/Production Cost Method

The area under the predicted LDC is the expected energy required by the system. If there were no generating units in the system, the expected energy not supplied (EENS) would therefore be equal to the area under the LDC. Let this be EENS(0).

The principle of the method, which is fully described in References 5 and 13, is to first decide the merit order of the units. The first unit in the merit order table is considered and its capacity outage probability table is evaluated. The energy not supplied (ENS) given the unit is in each of its possible states is evaluated by deducing the remaining area under the LDC. The EENS by the first unit is therefore the summation of the ENS by each of the unit states weighted by the probability of occurrence of each state. Let this be EENS(1).

The expected energy supplied by the first unit is therefore given by EENS(0) – EENS(1), and the expected production cost of the first unit will be this value of supplied energy multiplied by its individual production cost.

The method continues by including the second unit in the merit order table. A capacity outage probability

table consisting of both units is evaluated. The ENS by
each state of this new table is deduced and consequently
the EENS by the two unit system can be evaluated EENS(2).

The expected energy supplied by the second unit is
therefore EENS(1) - EENS(2). Its production cost can then
be evaluated.

This process continues until all the units have been
considered. The production cost of the system can be
evaluated by summating all the individual generating unit
production costs. The reliability of the system can be
assessed by one of several parameters. The first is the
EENS when all units have been considered. The second is
the "Energy Index of Unreliability (EIU)" or the "Energy
Index of Reliability (EIR)"; the EIR being the most usual
of the latter two. These are given by:

EIU = EENS/energy demanded,

EIR = 1 - EIU.

The Decision-Making Process

In practice the purpose of evaluating reliability indices and production costs is in order to aid the decision-making process. This must be based on parameters
which are quantitatively, objectively and consistently
evaluated. The method described in the previous section
is ideal in all these respects. However the decision
which has to be made is what units must be installed and
when. A generation expansion plan is therefore partly
decided on by considering alternative generation plans
and adding additional generating units to the model until
the reliability index (EENS/EIU/EIR) meets the level
which the utility management considers acceptable. In
addition, the type of units which are to be added depends
partly on the production cost that will be met by using
these units. In practice it may be decided to use a
large, cheaply operated unit (e.g. hydro or nuclear) high
up in the merit order and therefore low down in the LDC.

On the other hand it may be decided to use small but
expensive-to-operate units (e.g. gas turbines) as peaking
units and therefore low down in the merit order.

These decisions are never easy to make but it is certain that they cannot be made objectively without performing reliability and economic evaluations of the alternatives.

Application To The RTS

RTS Base Case. The technique described in the previous sections was applied to the RTS in Reference 12. The
following discussion reviews the results obtained. The
details and data for the RTS are fully documented in Appendix 1. The LDC (having a peak load of 2850 MW and a
total period of 364 days) is shown in Figure 2.1 and was
represented by 15 coordinate points. The merit order is
assumed to be in the order of increasing operating cost.

The area under the LDC of Figure 2.1 gives the value
of EENS(0) and is 14985204 MWh. This is the total energy
demanded and the energy not supplied if there are no generating units.

The capacity outage table of Unit 1 is shown in Table
2.9 together with the ENS given each of its states and
the EENS, i.e. EENS(1). The energy supplied by Unit 1 is
therefore EENS(0) - EENS(1), i.e. 432432 MWh. Since the
unit production cost of the hydro units are assumed to be
zero, the production cost at this point is also zero.

Table 2.9 - Capacity table and ENS for unit 1

Cap. out MW	Cap. in MW	Probability	ENS MWh	EENS MWh
0	50	0.99	14548404	4402919.96
50	0	0.01	14985204	149852.04
			EENS(1) =	4552772

The capacity outage table for Units 1 and 2 combined

is shown in Table 2.10 together with the ENS given each
of the states and the EENS, i.e. EENS(2).   The energy
supplied by Unit 2 is therefore EENS(1) - EENS(2), i.e.
432432 MWh.   This is the same as for Unit 1 since the
base of the LDC is still not completely full; the minimum
load being 998 MW.

Table 2.10 - Capacity table and ENS for units 1 and 2

Cap. out MW	Cap. in MW	Probability	ENS MWh	EENS MWh
0	100	0.9801	14111604	13830783.08
50	50	0.0198	14548404	288058.40
100	0	0.0001	14985204	1498.52
			EENS(2) =	14120340

This process of adding units is continued until all
units have been included.   The expected energy supplied
by each unit, its expected production cost and the overall values are shown in Table 2.11.   The production costs
given in Table 2.11 were calculated using the fuel costs
and heat rate values at maximum output given in Appendix
1 and Reference 6.   These figures show that the expected
energy not supplied by the system is 1320.5 MWh and that
the overall production cost is \$113.5 M.   The value of
EIU is 1320.5/14985204 = 0.000088 and the EIR = 0.999912.
This value of EENS could be reduced by including more
units but at the expense of increased production cost.
At this point, it relies on management to make the most
appropriate decisions.   These reliability indices evaluated for the RTS can be compared with the exact ones
given in the previous section, i.e. EENS (LOEE) = 1176
MWh and EIR = 0.999923.   The differences are negligible
and are due to approximations made in the representation
of the LDC.   The exact indices given before are therefore
very useful so that approximations such as these can be
assessed.

Table 2.11 – Energy and production costs for base case

Type	Capacity MW	Energy supplied GWh	Production cost/unit 1000$	Production cost/plant 1000$
hydro	50	432.43	0.0	
"	"	432.43	0.0	
"	"	432.43	0.0	
"	"	432.43	0.0	
"	"	432.43	0.0	
"	"	432.43	0.0	0.0
nuclear	400	3075.07	18450.42	
"	"	3060.86	18365.16	36815.58
coal 1	350	2528.82	28828.55	28828.55
coal 2	155	975.08	11349.93	
"	"	806.77	9390.80	
"	"	642.08	7473.81	
"	"	475.41	5533.77	33748.31
coal 3	76	182.21	2623.82	
"	"	149.41	2151.50	
"	"	119.49	1720.66	
"	"	94.91	1366.70	7862.68
oil 1	197	153.09	3383.29	
"	"	75.41	1666.56	
"	"	32.38	715.60	5765.45
oil 2	100	8.41	193.43	
"	"	5.10	117.30	
"	"	2.86	65.78	376.51
oil 3	12	0.25	6.90	
"	"	0.23	6.35	
"	"	0.22	6.07	
"	"	0.20	5.52	
"	"	0.19	5.24	30.08
gas turb.	20	0.26	11.31	
"	"	0.23	10.01	
"	"	0.20	8.70	
"	"	0.18	7.83	37.85
		14983.90	113465.01	113465.01

EENS (LOEE) of system = 1320.5 MWh

Total production cost = $113.5M

Effect Of System Load. The results shown in Tables
2.9 to 2.11 relate to a specific peak load of 2850 MW.
These results could be affected significantly by this
value of peak load. In order to illustrate this effect,
analyses were conducted [12] for peak loads ranging between ±7.5% of 2850 MW using the same generation system

and the same p.u. LDC shown in Figure 2.1. The results
of this analysis are shown in Figure 2.5.

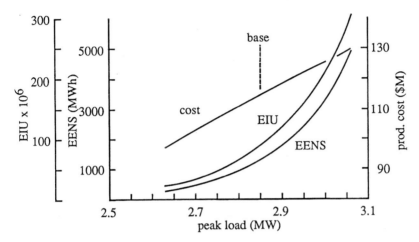

Figure 2.5 - Results for the RTS

     The results in Figure 2.5 clearly show that, as the
peak load increases without any increase in generating
capacity, i.e. the percentage reserve decreases, the
unreliability increases at an ever increasing rate. The
results also show that the shape of the EENS and the EIU
curves are very similar. This indicates that either of
the indices can be used to judge the system unreliability, that any trends shown by one would be reflected in
the other and that both respond to the system parameters
in a meaningful and objective manner.

     The production cost curve in Figure 2.5 is virtually
a straight line. Although the generating system remains
the same, the increased cost is due to the need to operate the more expensive plant for longer periods as the
peak load is increased.

     Effect Of Adding Different Sized Units. The most im-
portant requirement of a generation planning study is to
decide what units to add to the system and when in order
to meet the expected load growth.

In order to illustrate the effect of alternative expansion plans, consider the situation when the load increases by 7.5% to 3064 MW. The values of EENS and production cost for the base case (load = 2850 MW) and for
the increased load level are shown as cases (a) and (b)
in Table 2.12. It can be seen that, when the load increases, the EENS increases by almost a factor of 4 and
the production cost increases mainly due to the large
units operating for longer periods.

Table 2.12 – Effect of adding units

Case	Load MW	EENS MWh	Production cost, $M
(a) base case	2850	1320.5	113.46
(b) case case + 7.5% increased load	3064	4961.8	129.65
(c) (b) + 1x400 MW nuclear unit	3064	835.8	108.58
(d) (b) + 20x20 MW gas turbines	3064	390.8	129.85
(e) (b) + 14x20 MW gas turbines	3064	895.7	129.83

If the new EENS is considered unacceptable, additional units must be added. Assume that an additional
400 MW of capacity is considered appropriate. This capacity could be a single, low operating cost unit to be
operated as a base unit or several, high operating cost
units to be operated as peaking units. This is illustrated by cases (c) and (d) in Table 2.12 which shows the
effect of adding one 400 MW nuclear unit and 20 x 20 MW
gas turbines. It can be seen that the nuclear unit reduces the operating cost because it displaces some of the
energy produced by peaking units. On the other hand, the
EENS is greater than for the 20 gas turbines because the
probability of losing this one unit is very much greater
than that of losing the equivalent capacity of the gas
turbines.

It is evident therefore that it is not necessary to
add an identical capacity of gas turbines to achieve the
same EENS as that when the nuclear unit is added. In

fact, only 14 gas turbines need to be added to achieve
the same value of EENS. This is shown by case (e) in
Table 2.12. The production cost changes very little however since the incremental cost created by each gas turbine is very small compared with that of the large base
units.

These results clearly indicate that many alternatives
must be studied, in addition to those shown in this section, before a decision can be made. They also show that
the percentage reserve criterion is totally inadequate
and that reliability and economics must be considered together in order to enhance decision-making. Finally, the
results clearly demonstrate that objective decisions can
only be made if the reliability is quantitatively evaluated.

Summary

Reliability evaluation is an important aspect of any
generating system. Although there are a number of techniques available, many of them cannot be used directly in
the economic comparison of alternative generating expansion plans. This section has described one particular
technique based on energy evaluation that can be used in
economic comparison. This technique, which evaluates the
expected energy not supplied by a system and the energy
index of reliability, can be used to evaluate simultaneously the energy supplied by each unit of the system.
Hence it enables the production cost of each unit and of
the system to be evaluated.

The two parameters, reliability index and production
cost, are both important in the decision-making process
and therefore the particular technique described in this
section forms a very sound basis for providing the appropriate input data to this decision-making process. Energy based methods are also invaluable when assessing the
reliability of limited energy sources such as wind power.

This concept and application is described in the next
section.

RELIABILITY AND ECONOMIC ASSESSMENT OF NOVEL GENERATION

Background

There is considerable interest [14-19] in using novel
methods to generate electrical energy, including wind,
wave, tidal and solar energy sources. The main reason is
that conventional fuels are limited and expensive, whereas these other forms, generally known as renewables, are
limitless and cheap to operate although expensive in capital cost. Wind energy is potentially one of the most
promising and trial systems are in operation.

Information is needed, during the decision-making
process of the planning phase, regarding the impact and
benefit that these renewable sources will have on system
reliability and economics if included in the generating
system. This information would enable objective planning
decisions to be made regarding the relative merits of installing novel forms of generation or increased conventional generating capacity. Unfortunately, renewables
provide energy only intermittently, their operation cannot be scheduled in advance and their output cannot be
predicted accurately. Consequently, an equivalent amount
of conventional capacity must still be available to ensure that the demand will be supplied when the energy
from the renewable sources is not available; therefore,
renewables must be considered as energy-replacement
sources and not capacity-replacement sources. This leads
to one particular problem because most standard techniques for performing reliability and economic assessment
are capacity based, whereas energy-based techniques are
required.

This section, which is based on Reference 19, considers this problem and extends the existing loss-of-energy
(LOE) approach [5] described in the previous section to

incorporate wind-energy-conversion systems (WECS) in the evaluation of the static capacity generation reliability. These reliability results are then used to calculate the economic benefits of WECS. The techniques enable the impact and benefit of installing WECS into the system to be assessed during the planning phase, and so permits relative objective decisions to be made regarding the installation of WECS or additional conventional capacity. Although the approach is described and discussed in terms of WECS, it is applicable and amenable to systems containing other types of renewable energy sources. The approach is again applied to the RTS [6].

Concept Of Approaches

Capacity Based Approach. The importance of assessing the economic benefits obtained from renewable energy sources, when they are operated alongside conventional fossil and nuclear plants was recognized in Reference 18 which assumed that the benefits produced by the renewable sources caused a capacity replacement in the generating system. This capacity replacement was divided into two parts: the output of modern coal plants being displaced by the average power output of the renewables and the output of old and expensive plant being displaced by the firm power of the renewables. The firm power was defined as the proportion of renewable capacity which could be relied on to be available at times of peak demand.

This method is an approximation and is not consistent with actual operating behavior. The main disadvantages are:
(a) it assumes that the average power and firm power of the WECS are available on a continuous basis and can be specifically scheduled within the load duration curve,
(b) it assumes that all system units generate their average capacity continuously, instead of cycling

between full capacity, zero capacity and possibly
derated states, in a random manner,

(c) it does not assess system reliability and, therefore,
does not evaluate the technical impact of renewable
plant.

These disadvantages and weaknesses can be overcome by
using energy-based reliability techniques.

Energy Based Approach. This approach is based on the
loss-of-energy (LOE) technique described in the previous
section [5]. In this method, the generating system model
is combined with the load model to calculate the expected
energy not supplied (EENS) by the system due to unit
forced outages and the energy supplied by each individual
unit. In its basic form, the approach [5] assumes that
generating units are not energy limited other than by
their capacity and forced outage rate. The WECS and
other renewable sources, however, are energy limited by
virtue that they can only generate energy intermittently.
For this reason, they are treated [5,13,19] as negative
loads in order to modify the estimated load model. This
is performed using a conditional probability approach
[10]. The modified load model is then combined [5,13,19]
with the generation model using the basic LOE approach to
evaluate the system reliability indices.

The present economic approach [19] is based on the
assumption that the primary economic benefits of wind
plants are:

(a) to save fossil fuel in conventional plants,

(b) to contribute to system reliability.

These economic benefits are assessed in the present approach by using the energy reliability indices as input
parameters in the evaluation of production costs. This
approach, therefore, creates a combined reliability/eco-
nomic technique based on the assumption that WECS will
cause an energy replacement instead of capacity replace-

ment due to the intermittent nature of the wind; this being their primary source of energy.

Reliability Models

The objective of the reliability assessment is to
calculate the reliability index of the system and the expected energy supplied by each generating unit. The basic probabilistic approach for calculating these values
can be summarized by the following steps:
(a) select the appropriate model for each generating
    unit, depending on the characteristics of each unit,
(b) develop a capacity model for the generating system
    using the models of each unit,
(c) develop a suitable load model from the estimated data
    over the period of study,
(d) combine the capacity model with the load model to
    obtain the reliability index of the system and the
    expected energy supplied by each generating unit.
It is evident, therefore, that appropriate models for the
generation and for the load must first be established.

The generating units are divided into two groups:
the conventional units, which may be controlled and
scheduled, and the nonconventional units, i.e. WECS,
which generally cannot be scheduled. The conventional
units are represented by a two state model (up and down)
or by a multistate model that includes derated states.
The nonconventional units are represented by several
partial-output states, the number of states being dependent on the type of wind data available, the nature of
the wind regime, characteristics of the wind turbine,
availability of computational time and accuracy desired
[17].

The capacity model for the generating system is the
capacity outage probability table (COPT) [5]. The capacity models for the two groups, conventional and nonconventional, must be kept separate. The COPT for the con-

ventional units can be readily evaluated using standard
recursive techniques [5]. The COPT for the nonconventional units must also recognize the nature of the wind
data but can be evaluated using techniques given in
Reference 17. The load model required to evaluate the
reliability of generating systems is the Load Duration
Curve (LDC).

Reliability Assessment

The reliability and energy-production assessment
techniques are based on the loss-of-energy method described previously. This calculates the expected energy
not supplied (EENS) due to unit forced outages using:

$$\text{EENS} = \sum_{i=1}^{N} A_i \cdot P_i \qquad (2.4)$$

where N = total number of capacity states in the current
           system capacity outage probability table,

       A_i = area under load duration curve above a load
           equal to the capacity of the ith capacity
           state,

and P_i = probability of the ith capacity state.
The WECS provide electric energy only intermittently.
For this reason, they are treated as negative loads to
take into account their energy contribution to the system. This produces a modified load duration curve using
the techniques described below. This modified load duration curve is a model of the effective load which is perceived to exist and therefore to be satisfied by the remaining units in the system. It can be subsequently used
in the basic LOE approach to determine the system reliability.

The 'peak shaving' method [5,13] is used to modify
the load duration curve using a conditional probability
approach [10] and considering only the capacity probabil-

ity distribution of the WECS.

The 'peak shaving' technique [5,13] can be summarized by the following steps:

(a) compute the load points of the modified load duration curve, which are given by:

$$L_c(j) = L_o(k) - C(i) \qquad\qquad (2.5)$$

$$k = 1, \ldots, N_o$$

$$i = 1, \ldots, N$$

where $L_c(j)$ = load (jMW) of the modified load duration curve,

$L_o(j)$ = load (jMW) of the original load duration curve,

N_o = number of points of the original load duration curve,

N = number of capacity states of the wind plant,

and $C(i)$ = output capacity of ith capacity state of the wind plant.

If $L_c(j) < 0$, set $L_c(j) = 0$.

(b) compute the duration of load on the modified curve, using the following equation:

$$D[L_c(j)] = \sum_{i=1}^{N} d_i[L_c(j)] \times P_i \qquad\qquad (2.6)$$

where $D[L_c(j)]$ = duration of load $L_c(j)$ on the 'capacity modified curve',

P_i = probability of capacity $C(i)$,

and $d_i(L_c(j)]$ = duration of load $L_c(j)$ on the original load duration curve when reduced by $C(i)$.

The energy production model is summarized in the following steps:

(a) order the generating units according to their merit order, including the WECS,

(b) compute the total energy requirements of the system
during the period under study,

(c) compute the expected energy output of each generating
unit.

If the running cost of the WECS is less than that of the
other units in the system, then this plant should appear
first in the priority list and thus be loaded first. The
expected energy output of this plant is equal to the area
between the original load duration curve and the modified
load duration curve. The conventional units are then
added, one at a time, to the capacity outage probability
table. This table is combined with the modified load duration curve to determine the expected energy output of
each unit. The expected energy supplied by each conventional unit is equal to the difference between the expected energy not supplied before the unit is added and
the expected energy not supplied after the unit is added.

If the running cost of the WECS is not the lowest in
the system, the load duration curve is modified twice using the peak shaving technique in both stages of modification. In the first stage, the original load duration
curve is modified by the capacity outage probability
table of those conventional units placed at a higher priority in the merit order list. This modified curve is
the new original load duration curve. In the second
stage, this new original load duration curve is modified
by the capacity outage probability table of the WECS.

The expected energy output of the wind plant is equal
to the area between the new original load duration curve
and the modified new original load duration curve. The
expected energy output of the conventional units is calculated using the procedure described above. The original load duration curve and the modified new original
load duration curve are used to determine the expected
energy output of the conventional units with lower and
higher running costs than the wind plant running cost,

respectively.

Economic Assessment

When intermittent energy sources are assembled into the electricity supply on a large scale, the assessment should include, in addition to the system reliability evaluation, the costs and public acceptance of these energy sources. The capital cost per kilowatt installed of a WECS is still relatively large compared to that of nuclear or fossil plants. However, it is expected that these costs will decrease as mass production is introduced. A number of interesting developments in wind power have occurred in several countries. It seems that the capital cost per kilowatt of generating capacity for wind turbines, when built in large numbers, could be about the same as for a nuclear power station.

These and other considerations, including the shortage of operational experience with megawatt-sized machines, mean that capital costs are very uncertain. Therefore, as in Reference 18, it is better to calculate the break-even capital costs in order to determine the economic value of WECS.

The economic assessment is based on the assumption that the two major benefits of WECS to the generating system during its lifetime are:
(a) fuel savings,
(b) reduction in expected energy not supplied.

The former benefit is associated with the fact that the WECS energy will replace energy that would otherwise be supplied by plants having more expensive running costs.

The second benefit is associated with the fact that the impact of WECS on the system reliability has an improved physical meaning when it is expressed in terms of energy rather than power. Also, in countries very dependent on electrical energy, the value of reliability is

best measured by the enormous economic and social impacts
of interruptions. The magnitude of these impacts is highly dependent on the characteristics of the users (e.g.
type of user, size of operation, dependency) and on characteristics of the interruption (e.g. duration, frequency
and time of occurrence). This is discussed in Reference
20 and in Chapter 5. It is evident, however, that a real
measure of the interruptions is best given in terms of
energy.

The procedure for determining the WECS benefits is
summarized by the following steps:
(a) define a base case without WECS and calculate:
    (i) the expected energy output of each unit using
         the methodology described previously,
    (ii) the total system production cost which is equal
          to the sum of the expected energy supplied by
          each unit multiplied by its running cost per
          kilowatt-hour,
    (iii) the system reliability measured by the expected
           energy not supplied (EENS).
(b) add the WECS to the base case and calculate the same
    items as in step (a),
(c) compare the total system production cost of the base
    case with and without the WECS; the difference will
    be equal to the fuel savings due to the WECS,
(d) compare the EENS of the system with and without the
    WECS. The difference will be equal to the reduction
    in the expected energy not supplied due to the WECS.
    The system reliability benefit created by the WECS
    can be expressed in economic terms by multiplying the
    EENS reduction by the estimated cost of kilowatt-
    hours not supplied,
(e) calculate the total WECS benefit B which is given by
    the fuel savings S plus the EENS reduction cost EC:

$$B = S + EC \qquad\qquad\qquad (2.7)$$

The above procedure, (a) to (e), must be repeated for each year of the WECS lifetime. Each base case must include the capacity expansion program which is necessary to meet the peak demand.

The WECS break-even capital cost BC is given by the net present value of the benefits:

$$BC = \sum_{i=1}^{N} \frac{B_i}{(1 + DR)^i} \qquad (2.8)$$

where   N = WECS lifetime,

  B_i = benefits due to WECS in year i,

and   DR = discount rate.

Decisions and engineering judgments can then be made by comparing the WECS break-even capital cost with actual capital costs.

Application To The RTS

System Data And Assumptions. The algorithm previously described for assessing the economic value of WECS has been applied to the RTS containing a 5 MW wind plant. The following assumptions were made:

(a) the capacity levels and associated probabilities of each unit of the WECS are those shown in Table 2.13. These values are those published in Reference 17 and are the actual values recorded for a system in Hawaii. Owing to the lack of information, it was assumed that the probability distribution shown in Table 2.13, which is for a typical month, is the same for the complete period under study, i.e. 364 x 24 = 8736 hours. This is not true in practice because the wind probability distribution varies throughout the year. Also, such sets of data will be a major variable between different machines on different sites. However, this assumption does not override the main objective of this example which is to illustrate the

Table 2.13 - Output capacity table for WECS

Capacity in %	Capacity out %	Probability
100	0	0.8029630
84	16	0.0029630
70	30	0.0118519
56	44	0.0059259
42	58	0.0251852
30	70	0.0222222
20	80	0.0192593
8	92	0.0162963
0	100	0.0933332

techniques,

(b) the merit order of the WECS follows the hydro plants
and precedes the nuclear units,

(c) the LDC was represented by 15 coordinate points as
shown in Figure 2.1. From this curve the total energy requirements and the annual load factor of the
system are 14985204 MWh and 60.2%, respectively,

(d) at the time of commissioning the WECS, it was assumed
that the conventional plant running costs are those
given in Reference 6 and shown in Appendix 1. To
maintain consistency with previously published techniques and data, it was assumed that the annual running cost of the WECS was the same as that given in
Reference 18, i.e. equal to 3% of the break-even
capital cost,

(e) it was assumed that the running cost of the conventional plant would rise linearly over a period of 25
years (the WECS lifetime) to 1.5 times their values
at the WECS commissioning date,

(f) the discount rate was assumed to be 5%,

(g) the cost of energy not supplied was assumed to be
$0.50/kWh,

(h) since generation expansion is outside the scope of
the present study, it was assumed that:

(i) the plant mix would be the same over the WECS

lifetime,

(ii) the reduction in the energy per year required from all plants having higher running cost than the WECS would be the same as that obtained at the WECS commissioning date,

(iii) the reduction in the EENS each year due to WECS would be the same as that obtained at the WECS commission date.

The study was divided into two parts:

(a) WECS break-even capital cost calculation for the basic assumptions,

(b) sensitivity analysis of WECS break-even capital cost to changes in the basic assumptions.

Break-Even Capital Cost Calculation. The expected energy supplied by each plant in the system with and without the WECS, is shown in Table 2.14. The system production costs per year before and after the WECS is added, the fuel savings per year due to the WECS and their net present values over the lifetime of the WECS are also shown in Table 2.14.

The single 5 MW wind plant reduces the EENS from 1320.5 to 1281.1 MWh/yr, i.e. 39.4 MWh/yr or 2.98%. This reduction, however, is only a very small percentage (0.11%) of the total energy produced by the wind plant (36.67 GWh/yr); therefore, the main contribution of the wind plant is in energy replacement rather than improving the system reliability. The reduced EENS cost/year is:

$$EC = \$39.4 \times 0.5 \times 10^3 = \$19.7 \times 10^3$$

and its net present value NEC over 25 years is:

$$NEC = \$10^3 \times \sum_{i=1}^{25} 19.7/(1 + 0.05)^i = \$277.65 \times 10^3$$

The break-even capital cost of the WECS is equal to the sum of the net present values of its benefits, i.e. from NEC and the values given in Table 2.14:

Table 2.14 - Results for the RTS

Plant	Expected energy output		Reduction in expected energy output		Running costs		Fuel savings	
	without wind unit	with wind unit			without WECS	with WECS	per year	net present value
	GWh	GWh	GWh	%	$ x 10³	$ x 10³	$ x 10³	$ x 10³
Hydro	2594.59	2594.59	0	0	0	0	0	0
Wind	—	36.67	—	—	—	0.03 x BC*	-0.03 x BC*	-0.423 x BC*
Nuclear	6135.93	6134.71	1.21	0.02	36810	36802.70	7.30	122.49
Coal 1	2528.82	2523.22	5.60	0.02	28820	28756.20	63.80	1070.52
Coal 2	2899.34	2880.74	18.60	0.64	33750	33533.54	216.46	3631.92
Coal 3	546.01	539.46	6.55	1.20	7860	7765.69	94.31	1582.37
Oil 1	260.87	256.63	4.24	1.63	5770	5676.18	93.82	1574.23
Oil 2	16.37	16.00	0.37	2.28	380	371.42	8.58	143.97
Oil 3	1.08	.05	0.03	2.53	30	29.25	0.75	12.62
Gas turbine	0.87	0.85	0.02	2.69	40	38.98	1.02	17.14
Total	14983.88	14983.92	36.62	0.24	113460	112973.96 +0.03 x BC	486.04 -0.03 x BC	8155.26 -0.423 x BC

$$BC = \$(8155.26 - 0.423BC + 277.65) \times 10^3$$

thus

$$BC = \$5926.15 \times 10^3$$

Sensitivity Analyses.   These studies considered the parameters:

(a) fuel costs at WECS commissioning date,

(b) EENS cost,

(c) discount rate,

(d) load forecast uncertainty,

(e) number of plants.

Case (a).   The effect of fuel price on the break-even cost is depicted in Figure 2.6, which shows that it is more sensitive to the cost of coal.   The reason for this is that the expected energy output of the coal plants was reduced by a much greater amount due to the WECS than any of the other plant types.   This is indicated by the re-sults shown in Table 2.14, which shows that the reduction in coal-generated energy is 84% of the energy generated by the WECS.

This particular effect is very system specific and will depend greatly on the types of fuels that the WECS

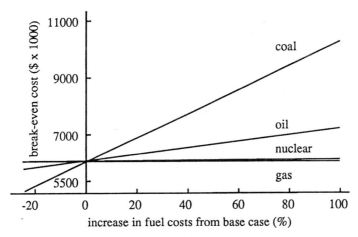

Figure 2.6 — Effect of fuel costs

displaces; therefore, the plant mix and fuel costs during
the lifetime of the WECS are two of the most important
parameters in assessing the economic value of the WECS.

Case (b). The effect of the cost of energy not supplied was found to be relatively small in the present example, as shown in Figure 2.7.   The reason is that the
present value of the fuel cost is far greater than that
of the energy-not-supplied costs, as shown previously.
The cost of energy not supplied would have to increase
very significantly to become an overriding factor in the
economics of the WECS.

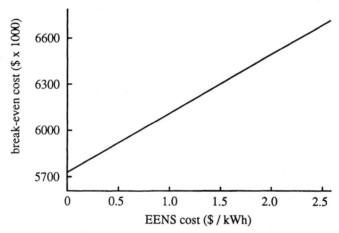

Figure 2.7 - Effect of EENS cost

Case (c). The effect of discount rate is shown in
Figure 2.8.   This shows that the break-even cost decreases significantly as the discount rate increases.

Case (d). The effect of uncertainty in the load demand is illustrated in Figure 2.9, which shows that the
break-even cost increases with demand.   The reason is
that running costs increase with load owing to more expensive plant being operated and for longer.   It becomes
more viable, therefore, to replace this high-cost energy
with WECS.   Also, the reduction in EENS increases with
increasing demand when WECS are installed. This increased

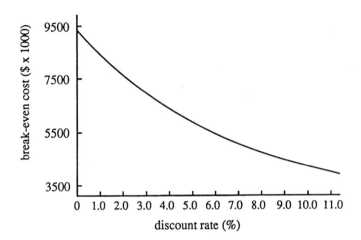

Figure 2.8 – Effect of discount rate

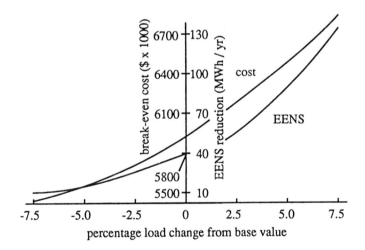

Figure 2.9 – Effect of load demand

reduction, which is also shown in Figure 2.9, enables higher break-even costs to become viable.

Case (e). In this case, a varying number of identical 5 MW wind energy plants were included. The effect on the break-even cost per plant and the EENS is shown in Figure 2.10.

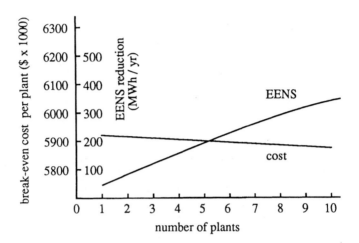

Figure 2.10 - Effect of WECS penetration

These results indicate that the break-even cost per
plant decreases slightly as the number of plants increases, i.e. as the penetration of WECS into the generation system increases.   The main reason is that the
reduction in EENS is nonlinear, as shown in Figure 2.10:
the decrement per plant decreasing as the number of
plants is increased.

Summary

Novel forms of generation are becoming of increasing
importance with wind energy conversion systems (WECS) being one of the most promising of these renewable energy
sources.   Owing to the intermittent nature of the wind,
the reliability and economic assessment of generating
systems containing WECS cannot be made using established
capacity-based methods.   This section has shown how energy-based reliability techniques can be used for these
assessments.   The main features of this approach are:
(a) it takes into account the intermittent nature of the
    wind energy,
(b) it takes into account the probabilistic nature of

forced outages,

(c) it takes into account fuel costs and plant mix in the system,

(d) it gives a measure of generating system reliability, and, therefore, is able to determine the impact of the WECS into the generating system,

(e) it is easy and simple to implement, which makes it a useful tool for assessing the long-term role of WECS,

(f) it enables the relative advantages of installing either WECS or additional conventional generation to be assessed objectively.

The numerical example indicates that:

(a) the WECS energy contribution has a greater impact on economics than on the EENS,

(b) the WECS break-even cost depends greatly upon the type of fuels that it displaces; therefore, the plant mix and fuel costs over its lifetime are two most important parameters in assessing the WECS economic value,

(c) the EENS reduction due to WECS becomes important when the cost of energy not supplied is very high compared with the conventional plant running costs,

(d) the WECS break-even cost tends to increase as the demand increases,

(e) the WECS break-even cost per plant tends to decrease as the WECS penetration in the system increases.

MONTE CARLO SIMULATION

General Concepts

As discussed in Chapter 1, there are two main reliability evaluation approaches; the analytical approach and the Monte Carlo Simulation (MCS) approach. This section discusses the MCS approach as an alternative to the analytical methods described in previous sections of this chapter.

The basic principle of MCS [21,22] is that it imi-

tates the operation of a system over a period of time.
It involves the generation of an artificial history of
the model of the system and the observation of that artificial history to draw inferences concerning the characteristics of the real system. This approach requires a
large amount of computing time and storage in order to
develop a good system model and, therefore, it should not
be used extensively if alternative analytical methods are
available. The simulation technique, however, is easy to
apply and can be used to solve not only simple problems
but also problems where direct analytical solutions may
not exist. Monte Carlo simulation is attractive because
of the flexibility it permits, as opposed to more restrictive analytical methods. In other words, the problem does not have to fit the model or technique; instead,
the model is developed to fit the problem.

Simulation techniques can be used to quantitatively
estimate the system reliability in even the most complex
system generating capacity situations [23,24]. Existing
methods for calculating generating system adequacy indices do not explicitly consider certain unit functions and
system operating policies. Monte Carlo simulation, however, provides a method of analysis which permits relax-
ation of many of the traditional assumptions incorporated
in the analytical techniques used to calculate adequacy
indices. It also provides a benchmark for comparison of
various modeling assumptions associated with analytical
techniques. Also a major shortcoming of most analytical
techniques is that they cannot provide the distributions
associated with the reliability indices. These distributions can be easily generated using simulation.

This section illustrates the utilization of simulation models to the RTS [6]. A range of sensitivity studies is conducted on the RTS to examine the effect of modeling assumptions and operating considerations on the
calculated adequacy indices.

Simulation Modeling

In simulation, the power system is modeled by speci-
fying a set of "events" where an "event" is a random or
deterministic occurrence that changes the "state" of the
system.   There are a number of ways in which the system
state can be defined.   The central measure used in this
chapter is the "available margin."   This is the difference between the available capacity (installed capacity
less failed units and capacity loss due to derating) and
load.   Planned outages are not considered but can be
easily incorporated.

The simulation model examines the system life during
a specified calendar year using repeated "yearly samples"
each consisting of 8736 hours (in the case of the RTS)
which are selected in their chronological succession (se-
quential approach).   This representation provides a convenient approach for including the system operational
characteristics (number of start-ups, shutdowns, etc.).
Additional advantage of using this approach is that it
can provide output in the form of probability distributions.

The advance of time is created by moving from one
event to another where each event creates new ones (e.g.,
a failure event creates a repair event, etc.)   The full
details of the models used in this section for the system
generating units, system load and system operating policies can be found in References [25-28] together with
details of the simulation processes used.

There are several adequacy indices that can be estimated from the simulated history of the system.   These
include:

(a) expected energy not supplied per interruption, (E,
    MWh/int),

(b) expected interruption duration, (D, hr/int),

(c) expected load loss per interruption, (ELL, MW/int),

(d) expected load loss per year, (ELL, MW/yr),

(e) loss of energy expectation, (LOEE, MWh/yr),

(f) loss of load expectation, (LOLE, hr/yr),

(g) frequency of interruptions, (F, int/yr).

These indices are cumulative, i.e., they represent all
prior years up to the current year of simulation. The
standard deviations of the adequacy indices can also be
obtained together with the distribution associated with
each adequacy index.

Convergence And Computing Time

It is difficult to make general statements regarding
the amount of simulation time required in order to have
reasonable confidence in the results of the simulation
model. However, experience with the model has indicated
that the simulation time required is only generally related to the system size. For example, the RTS [6] required a simulation time of only 35 to 40 years before
reasonable statistical convergence was observed in the
system adequacy indices. This system consists of 32 generators. However, when a test system consisting of only
11 generators was simulated, simulation runs approaching
60 years were necessary to achieve the desired level of
confidence in the system adequacy indices. In general,
it seems that homogeneous systems, i.e., systems in which
individual generators that are all relatively small with
respect to the total system capacity, converge readily
and quickly while systems not possessing this property
converge more slowly.

This convergence problem is illustrated for the RTS
base case in Figures 2.11 to 2.13 which show the oscillatory nature for the values of LOLE, F and D as the number
of simulated years is increased [27]. This oscillatory
pattern is a function of the initial seed and method used
to generate the random numbers. Therefore considerable
variations in the pattern can be expected.

The rate of convergence is not only affected by sys-

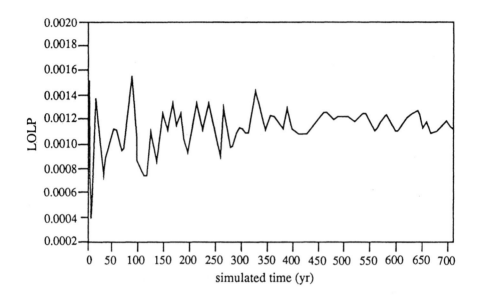

Figure 2.11 - LOLP of RTS as a function of simulated time

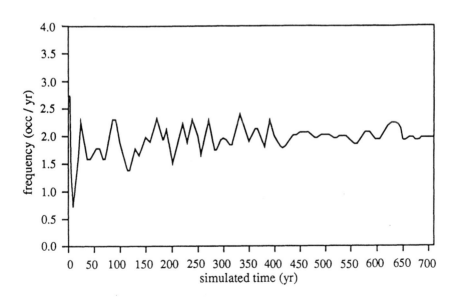

Figure 2.12 - Average capacity deficiency frequency of RTS as a function of simulated time

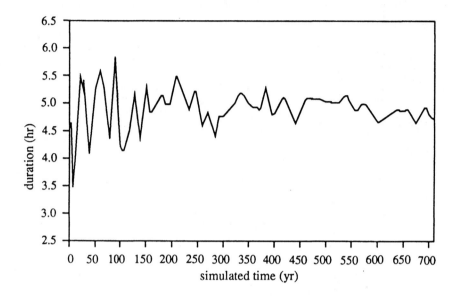

**Figure 2.13 – Average capacity deficiency duration of RTS
            as a function of simulated time**

tem size but also by the reliability of the units making
up a system. The number of simulated years to reach convergence increases as the unit reliabilities improve.
This is illustrated in Figures 2.14 and 2.15 which show
respectively the variation of LOLE for the RTS when the
MTTR of all units is doubled (decreased reliability) and
when the MTTR of all units is halved (increased reliability) [27].

The amount of required computing time constitutes the
only notable limitation of the simulation method. Systems with large numbers of generating units require more
computer time per simulated year than systems with a
smaller number of units. Since large systems, however,
require fewer simulated years for convergence, the computer time required for adequate convergence of the system adequacy indices is not entirely dependent on system
size.

An indication of the required computing time [27] is

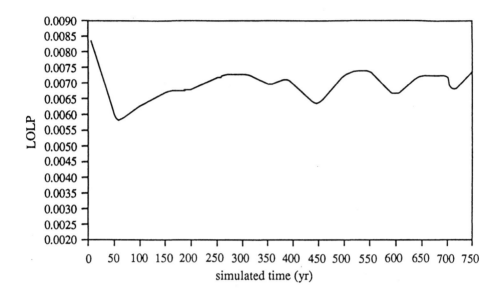

Figure 2.14 – LOLP of RTS – with doubled MTTR of all
units – as a function of simulated time

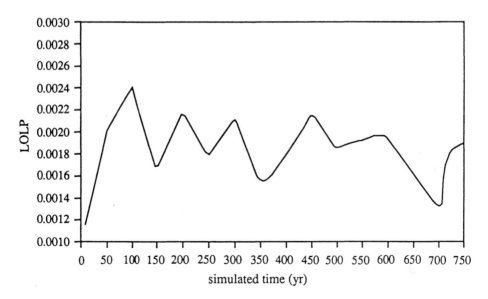

Figure 2.15 – LOLP of RTS – with halved MTTR of all units
– as a function of simulated time

shown in Table 2.15 for each simulated year of the RTS
base case.

Table 2.15 – Typical computing times for RTS

Computer	Processing	Time (sec)
CDC 7600	scalor	22.92
Cyber 205	scalor	2.90
Cyber 205	vector	0.26

It has also been reported [29] that an execution time
of 79 minutes was needed to simulate 600 years of an 18
generating system on an IBM 360/65.

These results clearly demonstrate the increasing
power of modern computers and the benefits that can be
accrued by vector or parallel processing. These improvements are making Monte Carlo simulation an increasingly
practical proposition and viable alternative to analytical techniques.

Results For RTS

This section illustrates some applications of the
model using the RTS. The generation system and load models for this system are given in Reference 6 and Appendi-
ces 1 and 2.

Case 1. This is the base case in which all units are
assumed to be running continuously except when in a state
of total outage. This basic approach is used in a conventional analytical study. The results are given in
Table 2.16. The LOEE and LOLE indices from Table 2.16
can be compared with those given in Reference 7 (LOEE =
1176 MWh/yr and LOLE = 9.39418 hr/yr) for a peak load of
2850 MW. There is a good agreement between the simulated
and analytical values. The difference is due to the limited simulation size.

The distributions of some of these adequacy indices
for a peak load of 2850 MW are shown in Figures 2.16 to

Table 2.16 - RTS adequacy indices for Case 1

	Peak load (MW)		
Index	2750	2850	2950
(a) Reference [25,26]			
E (MWh/int)	471.615	582.911	664.056
D (hr/int)	4.63726	4.85284	5.15625
ELL (MW/int)	71.4492	80.5525	84.8799
ELL (MW/yr)	73.4289	155.057	295.819
LOEE (MWh/yr)	484.683	1122.05	2314.33
LOLE (hr/yr)	4.76574	9.34130	17.9703
F (int/yr)	1.02771	1.92492	3.48515
Sim. time (yr)	397.0	293.0	202.0
(b) Reference [27]			
LOLE (hr/yr)		9.33632	
F (int/yr)		1.98592	
D (hr/int)		4.71418	
Sim. time		710.0	

2.19. These distributions are generally exponential in shape. The distributions of the load interruption duration and the frequency of interruptions are discrete.

Case 2. This is the same as Case 1, except that the 350 MW and the 400 MW units are assumed to have the three state model given in Appendix 2 [7]. The system peak load is 2850 MW. The adequacy indices are given in Table 2.17. It can be seen from this table that these indices decrease as more derated considerations are introduced.

Case 3. This is the same as Case 1 except that the 20 MW units are assumed to be peaking units [30]. The repair times for these units are assumed to be exponentially distributed. The system peak load is 2850 MW and the spinning reserve that the system must maintain is 0 MW. The random number sequence used in Case 1 is also employed in this case to ensure that the difference in the estimated adequacy indices is entirely due to the operating considerations. The adequacy indices are presented in Table 2.18. It can be seen that these indices are smaller than those of the base case (Case 1) for the same peak load. These indices approach the base case

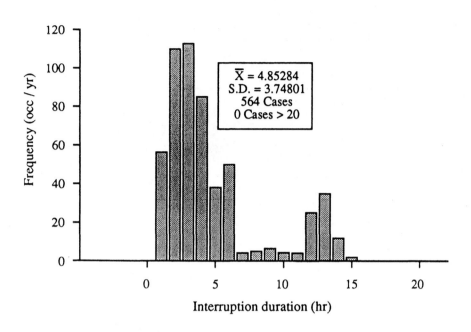

Figure 2.16 – Distribution of interruption duration

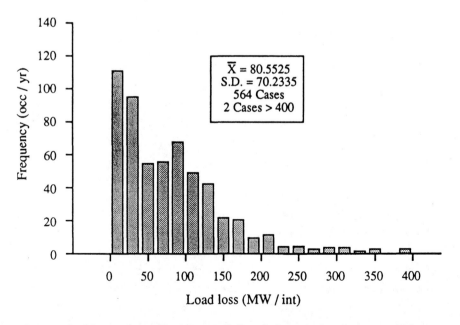

Figure 2.17 – Distribution of load loss per interruption

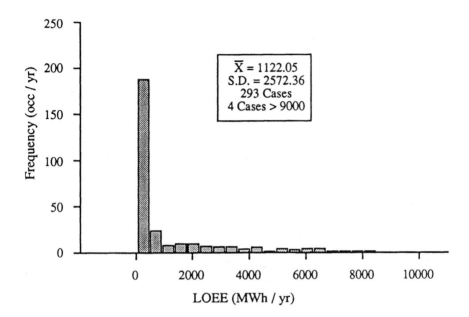

Figure 2.18 - Distribution of LOEE

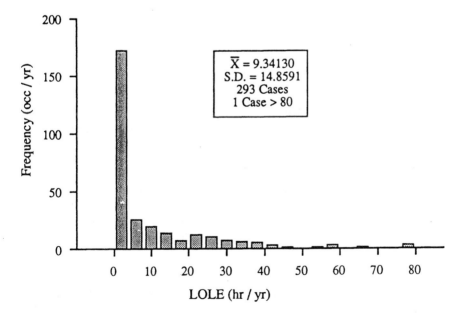

Figure 2.19 - Distribution of LOLE

Table 2.17 - RTS adequacy indices for Case 2

Index	Derated capacity (MW)			
	1x400	2x400	2x400 + 1x350	
E (MWh/int)	540.053	514.210	502.389	
D (hr/int)	4.63801	4.50424	4.34485	
ELL (MW/int)	77.7325	76.2869	75.4695	
ELL (MW/yr)	125.183	101.200	89.2688	LOEE
(MWh/yr)	869.718	682.139	594.246	
LOLE (hr/yr)	7.46933	5.97523	5.13909	
F (int/yr)	1.61043	1.32658	1.18284	
Sim. time (yr)	326.0	444.0	604.0	

Table 2.18 - RTS adequacy indices for Case 3

Index	Probability of start-up failure		
	0.00	0.01	0.05
E (MWh/int)	572.489	579.402	580.647
D (hr/int)	4.83742	4.84644	4.84588
ELL (MW/int)	80.6345	81.3132	81.3302
ELL (MW/yr)	145.307	148.195	154.888
LOEE (MWh/yr)	1031.65	1055.98	1105.81
LOLE (hr/yr)	8.71733	8.83277	9.22867
F (int/yr)	1.80205	1.82253	1.90444
Sim. time (yr)	293.0	293.0	293.0

values as the probability of start-up failure increases.

Case 4. This is the same as Case 3, except that the repair times of the 20 MW units are assumed to be Weibull distributed. The probability of start-up failure is 0.01. The adequacy indices are given in Table 2.19. It can be seen that the relative variations in these indices from the base values (b=1.0) are negligible due to the small amount of capacity allocated for the peaking operation. The effect of distributional assumptions will be much greater if a large percentage of the installed capacity is operated as peaking capacity.

Case 5. This is the same as Case 3 but with different spinning reserve margins. The probability of start-up failure is 0.01. The results are presented in Table 2.20. It can be seen that the adequacy indices increase

Table 2.19 - RTS adequacy indices for Case 4

Index	Distribution shape of repair time			
	β=0.5	β=1.0	β=2.0	β=4.0
E (MWh/int)	575.281	579.402	579.350	580.444
D (hr/int)	4.82804	4.84644	4.84486	4.85981
ELL (MW/int)	80.6676	81.3132	81.2670	81.1243
ELL (MW/yr)	147.294	148.195	148.388	148.128
LOEE (MWh/yr)	1050.43	1055.98	1057.85	1059.96
LOLE (hr/yr)	8.81570	8.83277	8.84642	8.87372
F (int/yr)	1.82594	1.82253	1.82594	1.82594
Sim. time (yr)	293.0	293.0	293.0	293.0

as the required spinning reserve margin increases. It is
expected that for higher reserve margins the adequacy
indices will increase very sharply because the peaking
units will be used to satisfy the reserve requirements.
Thus, it is likely that these units will not be available
when needed to satisfy the load demand.

Table 2.20 - RTS adequacy indices for Case 5

Index	Spinning reserve margin (MW)		
	50	100	400
E (MWh/int)	580.332	580.359	581.936
D (hr/int)	4.85335	4.85493	4.88141
ELL (MW/int)	81.3600	81.4734	81.7555
ELL (MW/yr)	149.100	150.174	151.706
LOEE (MWh/yr)	1063.51	1069.74	1079.85
LOLE (hr/yr)	8.89420	8.94881	9.05800
F (int/yr)	1.83259	1.84324	1.85561
Sim. time (yr)	293.0	293.0	293.0

Summary

    This section has illustrated that Monte Carlo simula-
tion is a practical method for assessing generating system adequacy. This approach makes the examination of
complex systems possible without forcing the system model
to become unrealistic. It also constitutes a tool for
allowing easy modification of the number and characteristics of input random quantities. It offers the planning

engineer a synthesis of the final results and a detailed
description of the events that caused the results. The
latter output provides operating experience which can facilitate the exchange of ideas between the planning engineer and the operating engineer.

The application of Monte Carlo methods generally requires considerable computing time in order to obtain
sufficient confidence in the results. It should be recognized, however, that Monte Carlo simulation can be used
for detailed studies of specific configurations while
other techniques based on methods which are faster but
less general can be used for exploratory or expansion
studies.

It is worth noting that mixed analytical/Monte Carlo
simulation approaches have been used in order to take advantage of both. Reference [31] describes such an approach to the loss of load probability of hydrothermal
generating systems. The classical generation reliability
methods originally developed for systems with a high proportion of thermal units are inadequate for such systems,
because they assume that the generating capacity depends
only on forced outages [5]. It has been shown that the
equipment outages have a very small effect on the energy
state of a large hydroelectric system and the reservoir
depletion on the output capacities of the units pre-
dominates. The depletion and outages have been decoupled. The complex distribution of reservoir depletion
has been represented by simulation, while the equipment
outages have been dealt with by analytical methods [31].

REFERENCES

1. Billinton, R., "Bibliography On Application Of Probability Methods In The Evaluation Of Generating Capacity Requirements," IEEE Winter Power Meeting,
   1966, paper 31 CP 66-62.
2. Billinton, R., "Bibliography On The Application Of
   Probability Methods In Power System Reliability
   Evaluation," IEEE Trans. on Power Apparatus and Systems, PAS-91, 1972, pp. 649-660.

3. IEEE Committee Report, "Bibliography On The Application Of Probability Methods In Power System Reliability Evaluation, 1971-1977," IEEE Trans. on Power Apparatus and Systems, PAS-97, 1978, pp. 2235-2242.
4. Allan, R.N., Billinton, R. and Lee, S.H., "Bibliography On The Application Of Probability Methods In Power System Reliability Evaluation, 1977-1982," IEEE Trans on Power Apparatus and Systems, PAS-103, 1984, pp. 275-282.
5. Billinton, R. and Allan, R.N., "Reliability Evaluation Of Power Systems," Longman, London, (England)/ Plenum Press, New York, 1984.
6. IEEE Committee Report, "IEEE Reliability Test System," IEEE Trans. on Power Apparatus and Systems, PAS-98, 1979, pp. 2047-2054.
7. Allan, R.N., Billinton, R. and Abdel-Gawad, N.M., "The IEEE Reliability Test System - Extensions To And Evaluation Of The Generating System," IEEE Trans on Power Systems, PWRS-1, No. 4, 1986, pp. 1-7.
8. Rau, N.S. and Schenk, K.F., "Application Of Fourier Methods For The Evaluation Of Capacity Outage Probabilities," IEEE Winter Power Meeting, 1979, New York, paper A79 103-3.
9. IEEE Std 762, "Definitions For Use In Reporting Electric Generating Unit Reliability, Availability And Productivity".
10. Billinton, R. and Allan, R.N., "Reliability Evaluation Of Engineering Systems, Concepts And Techniques," Longman, London, (England)/Plenum Press, New York, 1983.
11. Billinton, R. and El-Sheikhi, F.A., "Preventive Maintenance Scheduling Of Generating Units In Interconnected Systems," International RAM Conference, 1983, pp. 364-370.
12. Allan, R.N. and Corredor-Avella, P., "Energy Based Indices In The Reliability Evaluation Of Generating Systems," Inter Conf STAQUAREL '84, Prague, March 1984, pp. 11-20.
13. Billinton, R. and Harrington, P.G., "Reliability Evaluation In Energy Limited Generating Capacity Studies," IEEE Trans. on Power Apparatus and Systems, PAS-97, 1978, pp. 2076-2085.
14. British Wind Energy Association, "Wind Energy For The Eighties," (Peter Peregrinus, 1982).
15. Milborrow, D.J., "Wind Power In The UK Electricity Supply Industry," Electron. & Power, 1982, 28, pp. 665-669.
16. Taylor, D., "Renewables Prospects For Britain's Utilities," Electr. Rev., 1983, 212, pp. 24-26.
17. Desmukh, R.G. and Rama Kumar, R., "Reliability Analysis Of Combined Wind-Electric And Conventional Systems," Sol. Energy, 1982, 28, pp. 345-352.
18. Cottrill, J.E.J., "Economic Assessment Of The Renewable Energy Sources," IEE Proc. A, 1980, 127, (5), pp. 279-288.

19. Allan, R.N. and Corredor-Avella, P., "Reliability And
    Economic Assessment Of Generating Systems Containing
    Wind Energy Sources," Proc. IEE, 132 Pt. C, 1985, pp.
    8-13.
20. Billinton, R., Wacker, G. and Wojczynski, E., "Cus-
    tomer Damage Resulting From Electric Service Inter-
    ruptions," Canadian Electrical Association Report,
    April 1982.
21. Hammersley, J.M. and Handscomb, D.C., "Monte Carlo
    Methods," John Wiley & Sons, Inc., New York, 1964.
22. Rubinstein, R.Y., "Simulation And The Monte Carlo
    Method," John Wiley and Sons, New York, 1981.
23. Noferi, P.L., Paris, L. and Salvaderi, L., "Monte
    Carlo Methods For Power System Evaluation In Trans-
    mission Of Generation Planning," Proceedings 1975 An-
    nual Reliability and Maintainability Symposium, Wash-
    ington, 1975.
24. EPRI Report, "Modeling Of Unit Operating Consider-
    ations In Generating Capacity Reliability Evaluation.
    Volume 1: Mathematical Models, Computing Methods,
    And Results," Report EL-2519, Electric Power Research
    Institute, Palo Alto, Ca., July, 1982.
25. Ghajar, R., "Utilization Of Monte Carlo Simulation In
    Generating Capacity Planning," M.Sc. Thesis, College
    of Graduate Studies and Research, University of
    Saskatchewan, Saskatoon, September, 1986.
26. Billinton, R. and Ghajar, R., "Utilization Of Monte
    Carlo Simulation In Generating Capacity Adequacy
    Evaluation," CEA Transactions, Volume 26, 1987.
27. Jebril, Y.A.A., "Monte Carlo Simulation In Power Sys-
    tem Reliability Evaluation," M.Sc. Dissertation,
    UMIST, 1985.
28. Saboury, A., "Monte Carlo Methods In Reliability
    Evaluation Of Hydrothermal Generating Systems," M.Sc.
    Dissertation, UMIST, 1986.
29. Ayoub, A.K., Guy, J.D. and Patton, A.D., "Evaluation
    and Comparison Of Some Methods For Calculating Gener-
    ating System Reliability," IEEE Trans. on Power Appa-
    ratus and Systems, PAS-89, 1970, pp. 537-544.
30. IEEE Task Group, "A Four-State Model For Estimation
    Of Outage Risk For Units In Peaking Service," IEEE
    Trans. on Power Apparatus and Systems, PAS-91, 1972,
    pp. 618-627.
31. Cunha, S.H.F., Gomes, F.B.M., Oliviera, G.C. and
    Pereira, M.V.F., "Reliability Evaluation In Hydro-
    thermal Generating Systems," IEEE Summer Power
    Meeting, 1982, paper 82 SM430-7.
32. Allan, R.N., Billinton, R., Shahidehpour, S.M. and
    Singh, C., "Bibliography On The Application Of Prob-
    ability Methods In Power System Reliability Evalua-
    tion, 1982-87," IEEE Winter Power Meeting, New York,
    February 1988.

CHAPTER 3

COMPOSITE SYSTEM ADEQUACY EVALUATION

INTRODUCTION

Power system planners, designers and operators have always been and continue to be extremely concerned about system reliability. This concern manifests itself in a wide array of qualitative and quantitative criteria for subsystem and system reliability assessment. The bulk of the published material pertains to Hierarchical Level I evaluation, where the basic concern is the ability of the generation system to generate sufficient energy to satisfy the system load requirement. The joint or composite generation and transmission problem of assessing the ability of the system to satisfy the load and energy requirements at the major load points is designated as Hierarchical Level II assessment. The concept of hierarchical levels is discussed in detail in Chapter 1. It can be seen in the IEEE PAS Bibliographies on power system reliability evaluation [1,2,3,60] that there is a much smaller body of literature available in the HLII area and the initiation of this activity came several decades after significant developments in HLI evaluation. This can clearly be seen from the references on the subject of HLII evaluation given at the end of this chapter. This evaluation domain involves the joint reliability problem of generating sources and transmission facilities and is sometimes called bulk system analysis.

HISTORICAL DEVELOPMENTS

Reliability evaluation of bulk generation and transmission systems came under active investigation in Europe
and North America in the 1960's. An early example of
joint consideration of generation and transmission facilities is shown in Reference 4. The term "composite system reliability evaluation", however, first appeared in
1969 [5,52]. The basic objective was to assess the
ability of the system to satisfy the real and reactive
power requirements at each major load point within acceptable voltage levels.

Two concurrent and independent streams of activity in
regard to composite system reliability evaluation appear
to have been initiated in Europe and North America during
the late 1960's [5-10]. These approaches to the assessment of composite system reliability are fundamentally
different and with subsequent development have become
known as the simulation and contingency enumeration methods, respectively. It is possible that the requirements
for modeling generating capacity in HLI evaluation played
a key role in selecting a suitable approach to composite
system reliability evaluation. The French and Italian
systems with significant hydro facilities including
pumped storage were strongly motivated to develop a
method capable of modeling hydro resources and therefore
utilized Monte Carlo or simulation methods. Precedents
derived from North American generating capacity reliability models tended to influence the North American approach to HLII assessment. Further work in both the areas of simulation [11-14] and contingency enumeration
[15,16,18] were reported in the early 1970's.

The contingency enumeration approach was extended by
developments outlined in References 16 and 18. Reference
16 can be considered as the initial publication in the
sequence of activity described in the EPRI Report on RP
1530 [50]. Reference 18 was the initial publication in

the activity which resulted in the GATOR program developed by Florida Power Corporation. An IEEE Working Group
identified in 1972 a number of indices which could be
used to assess bulk system reliability and published
their findings in Reference 17.

The general area of power system reliability evaluation and particularly questions regarding models and philosophy received considerable impetus at an 1978 EPRI
Workshop entitled "Power System Reliability - Research
Needs and Priorities". The proceedings [56] for this
conference contain many models and concepts which if not
directly related are important contributions to HLI and
HLII evaluation. Reference 19 is a direct application to
the area of composite system reliability evaluation.

A number of papers were presented in 1978 and 1979
which provided useful development work on composite system reliability assessment [18-26]. A comprehensive set
of indices for adequacy assessment is presented in Reference 24 and is further amplified in Reference 54. References 23 and 33 introduced a somewhat different approach
to composite system reliability evaluation by defining
the load supplying capability of an area or system. The
influence of common mode outages in HLII evaluation was
illustrated in 1981 in Reference 30 The consideration of
station related events was introduced by Reference 32 in
1981 and further extended in References 38, 39, 43 and
57. Further development in the contingency enumeration
approach was reported in References 34 and 41 and introduced the concept of placing upper and lower bounds on
the calculated system indices [41].

The IEEE Power Engineering Society presented a panel
discussion at the 1983 Winter Power Meeting on the subject of HLII assessment and subsequently published two
papers arising from this activity. These papers [35,36]
provided a timely reference on a number of different
viewpoints. A related topic which might be considered to

lie somewhere between HLI and HLII evaluation is that of
transfer capability between two areas. Incorporation of
probabilistic considerations in transfer capability is
described in Reference 37 and further examined in Reference 45.

A comparative study of existing digital computer programs for composite system adequacy evaluations was conducted on behalf of the Canadian Electrical Association
Power System Reliability Subsection. The results of this
study are detailed in Reference 42 and provide an interesting illustration of the different perceptions and
therefore objectives in composite system adequacy assessment. A comparison between a contingency enumeration approach and a simulation technique was published in 1985
using the IEEE Reliability Test System (RTS) [58] and is
described in Reference 46. The simulation approach which
had previously been advanced by EDF (France) and ENEL
(Italy) was further illustrated by work in Brazil which
is documented in Reference 44. Recent work on the recognition of load point indices and an examination of pertinent factors in composite system adequacy evaluation are
described in References 47 and 48. Work done on applications for large system analysis was reported in Reference
49.

The bulk of the references given at the end of this
chapter refer to conceptual development work or to applications to relatively small systems such as the RTS. They
incorporate a wide range of system concepts from the ac
load flow representation used in some contingency enumeration methods to transportation or dc load flow techniques generally used in the simulation applications.
Reference 42 provides a comparison between the capabili-
ties of the basic contingency enumeration programs presently available together with an indication of their
limitations. References 50 and 51 provide detailed information on direct EPRI related research in HLII assess-

ment. Reference 50 describes the prototype program SYREL
developed by Power Technologies Inc. under EPRI sponsor-
ship while Reference 51 is a detailed examination of
overall adequacy indices. The work described in Reference 50 is being extended to the development of functional specifications for a digital computer program
suitable for large system adequacy assessment. There are
many additional publications which indirectly relate to
the subject of HLII or composite system reliability evaluation. Publications on generation and transmission line
reliability data and models all relate in some way to
HLII adequacy assessment i.e. the ability to supply the
required power and energy at the major system load
points. The references at the end of this chapter have
been constrained by including only those publications
which are believed to directly relate to HLII assessment
and which clearly illustrate the evolution of the area to
its present status.

Adequacy assessment at the HLII is a complex task and
there is no single accepted procedure. As noted in this
section, there are two basic techniques, each of which
has many variations. The following sections examine some
of the pertinent factors in the contingency enumeration
approach and present a comparison between a contingency
enumeration analysis and a Monte Carlo simulation study
of the RTS.

FACTORS IN THE CONTINGENCY ENUMERATION APPROACH

Background

The quantitative evaluation of the adequacy of a composite power system normally comprises the following basic steps:
(a) evaluate the performance of the power system without
    removing any component, or in other words study the
    performance of the base case system,
(b) make changes in the network configuration due to the

"credible" outage(s) of various components,

(c) check the adequacy of the modified power system,

(d) take any correction action, if necessary, such as
    rescheduling of the generating units, line overloads
    alleviation, correction of bus voltages and load
    curtailment at buses, etc.,

(e) calculate the adequacy indices for the system as a
    whole and the individual load points.

The entire process is quite complex and the computation time required to analyze a practical network is very
large.   There is a wide variation both in terms of techniques utilized to evaluate the adequacy of a power system, and the quantitative indices created to reflect the
adequacy of the entire system and the individual load
centers [42]. Two sets of indices; individual load point
indices and system indices are required in order to obtain a complete assessment of a bulk transmission system.
It has been emphasized in a recent paper [47] that the
two sets of indices respond quite differently to variations in system parameters.   It is, therefore, quite
important in the adequacy assessment of a power network
to recognize those factors that are pertinent to the adequacy of the system and/or major load centers.   These
factors, which were discussed in detail in Reference 48,
are influenced by the size of the system, the network
topology and the intent behind the adequacy studies.
Some of the factors whose effects are quite pronounced in
a composite generation and transmission system adequacy
evaluation are:

(a) selection of an appropriate network solution tech-
    nique,

(b) selection of an appropriate load curtailment philoso-
    phy,

(c) selection of an appropriate contingency level both
    for generating unit and transmission line outages,

(d) inclusion of station originated and common cause

events.

The effects of these factors on the adequacy indices are illustrated in the following sections using the RTS.

Appropriate Network Solution Techniques

The adequacy assessment of a bulk power system involves the solution of a network configuration under selected outage situations. Various solution techniques, depending upon the adequacy criteria employed and the intent behind the studies are available in order to analyze the adequacy of a power system. Transportation models, d.c. load flow and a.c. load flow methods are the most commonly used techniques. When both continuity and quality of the power supply are of concern, then it is necessary to examine the voltage levels at each major load center and the VAr limits of each generating unit while considering the effect of component outages i.e. generating units, transmission lines and transformers. Considering a power network as a transportation model or using approximate simple load flow techniques such as d.c. load flow etc. does not provide an estimate of the bus voltages and the reactive power limits of the generating units. If the quality of power supply is an important adequacy criterion, then more accurate a.c. load flow methods such as Gauss-Seidel, Newton-Raphson and second order load flow techniques must be employed to assess the adequacy. These techniques are computationally more expensive than approximate methods and have large storage requirements. In these cases, it is therefore mandatory in large system studies to use computationally fast a.c. load flow techniques. Digital computer programs using the fast decoupled load flow technique has been developed independently at the University of Saskatchewan and UMIST. This load flow technique has proved to be very satisfactory for most situations.

In some outage situations, the load flow solution may

fail to converge. In this case, a quantitative assessment of the situation becomes extremely difficult. These
non-convergence situations are frequently encountered
when considering the outages of transmission lines and
transformers. The simplest solution to this situation is
to allow the system buses to stay at the low voltage and
treat the outage event as a system failure due to the
voltage violation(s). This approach, however does not
provide a quantitative measure of the voltage violation
problem and also does not give due consideration to the
severity of the outage event. These events are treated
as failure events regardless of the voltage magnitude at
the system buses. A major objection to this approach is
whether low voltage really constitutes a system/bus failure. Many power utilities use d.c. load flow for adequacy
studies because they do not view low bus voltage as a
failure but only as a minor problem which is normally
rectified by transformer tap-settings and/or phase-shift-
er adjustments. This treatment, however, gives an optimistic assessment because correction of voltage violations may not be possible for all voltage violation contingencies. It also does not permit a quantitative
evaluation of the outage contingencies using voltage as
an adequacy criterion. The actual situation lies somewhere between the two viewpoints. It is desirable to use
voltage as an adequacy criterion but suitable corrective
action should be taken when encountering any voltage
problem.

Most non-convergent situations are due to high values
of the mismatch in reactive power beyond the permissible
tolerance limit. Very few situations are due to high
values of the mismatch in active power. Another possibility is that a load flow may not converge although a
solution, in fact, does exist. This could occur due to
numerical problems with the fast decoupled algorithm
and/or the characteristics of the numerical formulations

used. These problems arising with an a.c. load flow can, however, be treated by employing suitable corrective actions and it should be emphasized that these problems are not major obstacles to employing a.c. load flow techniques in adequacy evaluation of a power network. In certain extreme cases, it may be appropriate to utilize an approximate technique such as dc load flow or a transportation model approach to solve a particular contingency case.

Appropriate Load Curtailment Philosophies

Concepts. An important consideration in composite system adequacy evaluation is the curtailment of load at the appropriate buses under a capacity deficiency in the system [48]. The curtailment of load at system buses in the event of a deficiency in the generation capacity can occur in a number of ways depending upon the relative priority given to each major load center. The effect of load curtailment philosophy on adequacy indices can be illustrated as follows:

Assume that the load at each bus is classified into two categories:
(a) firm load,
(b) curtailable load.

Based on individual load point requirements, curtailable load may represent some percentage of the total load at the bus. In the case of a deficiency in the generation capacity, curtailable load is interrupted first, followed by the curtailment of firm load, if necessary. The effect of a system disturbance that results in swing bus overload, (a capacity deficiency in the system) can be confined to a small area or to a large region of the system. If the relative importance of the load at a bus in the system is such that the firm load at the bus will not be curtailed unless it is unavoidable, it is obvious that more buses in the system will experience load curtail-

ment.   On the other hand, if the system design warrants
that a disturbance in one region should not be felt in
another region of the system, then the number of buses
that experience load curtailment will be less, but under
many outage contingencies firm load may have to be curtailed.   This provision can be incorporated in the load
curtailment philosophy algorithm by defining the number
of load curtailment passes as follows:

Load Curtailment Pass 1.   In the case of generator
outages, pass 1 covers those buses at which the generators under outage are physically connected or are one
line away and receiving power from these generator buses.
In the case of line outages, the receiving end bus(es) of
lines under outage and buses which are one line away and
receiving power from these receiving end buses may experience load curtailment.   In the event of both generator
and line outages, the buses considered are those at which
the generator under outage is physically connected and
the receiving end bus of the line under outage together
with those buses which are one line away from the receiving end buses and are receiving power from them.   The
swing bus overload is alleviated by proportional interruption of the curtailable load at buses covered under
pass 1.   If the swing bus is still overloaded after removing the curtailable load from the buses mentioned earlier, the firm load is curtailed proportionally at these
buses.   However, at those buses which have load as well
as local generation, only that amount of firm load is
curtailed which is in excess of its local generation.   In
other words, these buses do not experience any firm load
curtailment if the generation is more than the firm load.
If generation is less than the firm load, the excess firm
load is interrupted proportionately.   If the swing bus is
overloaded even after curtailing the total load at the
load buses, the load is removed from those buses which

are covered under load curtailment pass 2 which is described as follows:

Load Curtailment Pass 2. In load curtailment pass 2,
the buses covered are as noted for pass 1 and all those
buses which are two lines away from the generator outage
buses and/or receiving end buses for a line outage and
are being directly supplied from the buses covered under
pass 1. The load curtailment philosophy remains the same
as described above, i.e. proportional curtailment of the
curtailable load followed by proportional curtailment of
the firm load, if necessary. If the swing bus is still
overloaded after removing the total load from the buses
covered under pass 2, the load is curtailed at the other
buses covered under pass 3.

Load Curtailment Pass 3. This pass covers all buses
that are covered under pass 2 and those additional buses
which are three lines away from the generator outage
buses and/or receiving end buses for a line outage and
are being fed from the buses which are two lines away and
covered under pass 2. The load curtailment philosophy
remains the same as for pass 1. If the swing bus is
still overloaded after curtailing the total load at all
buses covered under pass 3, the load is curtailed proportionately at all system buses. This possibility is, however, very remote. The principle can be used to extend
the number of load curtailment passes if required.

As noted earlier, the number of the buses at which
curtailable load is to be interrupted increases as the
number of load curtailment passes increases. The number
of passes can be specified depending upon the system requirements and the operating philosophy. The effect of
the number of load curtailment passes on the adequacy indices is discussed in the next section.

Effect Of Load Curtailment Passes

The single line diagram for the RTS and the system
data are given in Appendix 1. In this system, there are
17 buses which have loads connected at them. The remaining buses are either free buses or generator buses without connected load. Adequacy indices have been calculated for these 17 buses. The system load is assumed to be
2850 MW. The distribution of load at each bus is given
in Appendix 1. It has been assumed that the firm load at
a bus is 20% of the total load. In order to facilitate a
better comparison of the adequacy indices for these buses, they are classified into 6 categories depending upon
their type, voltage level and location relative to a generating station. This classification helps not only in
comparing the adequacy indices of buses falling into one
class with the adequacy indices of buses falling into
other classes, but also in achieving a better pictorial
representation of the adequacy indices. The buses in the
six categories are as follows:
(a) 138 kV buses (South Region):
    (i) buses having local generation: Buses 1, 2 and 7,
    (ii) one line away buses from generating stations and
         connected with two lines: Buses 4, 5 and 6,
    (iii) one line away buses from generating stations and
          terminated with three or more lines: Buses 3 and
          8,
    (iv) two lines away buses from generating stations:
         Buses 9 and 10,
(b) 230 kV Buses (North Region):
    (v) buses having local generation: Buses 13, 15, 16
        and 18,
    (vi) buses having no local generation: Buses 14, 19
         and 20.
On the basis of the above classification, the variation in the adequacy indices i.e. the probability of
failure, the frequency of failure and the expected load

curtailed in MW as a function of the load curtailment
passes are shown in Figures 3.1, 3.2 and 3.3 respectively. There are six sets of graphs in each figure.
The number of graphs in each set is not equal but depends
upon the number of buses in each class. The scale on
both the horizontal and vertical axes are the same for
all the six sets of graphs in each figure. This fa-
cilitates a quick comparison of the adequacy indices of
buses in one class to those of buses in another class.

It can be seen from Figures 3.1, 3.2, and 3.3 that
the indices for buses having local generation do not
change appreciably as the number of passes increases.
The most notable increase in indices is for buses 9 and
10 as the number of load curtailment passes increases
from 1 to 2. These buses are two lines away from generator bus 13 and bus 23. An outage of a 197 MW unit at
bus 13, or the outage of a unit at bus 23 with the outage
of other large generating units therefore results in load
interruption at these two buses. The net effect is that
these two buses also share the capacity deficiency. This
reduces the amount of load curtailment and also energy
curtailment at buses falling in the 230 kV region with
the exception of bus 19. In fact, bus 19 shares more of
a capacity deficiency at pass 2, because it is two lines
away from bus 23 which has large generating units (1 unit
of 350 MW and 2 units of 155 MW each).

One interesting feature is that the expected load
curtailed at buses 4, 6 and 8 decreases as the load curtailment pass is increased from 1 to 2, but if the load
curtailment pass is further increased from 2 to 3, the
expected load curtailed increases. This can be seen in
Table 3.1 which gives the expected load curtailed in MW
at each bus for the three passes and associated variation
(in p.u.) as a function of the expected load curtailed
for pass 1. This is due to the fact that, at load curtailment pass 2, buses 9 and 10 share the load curtail-

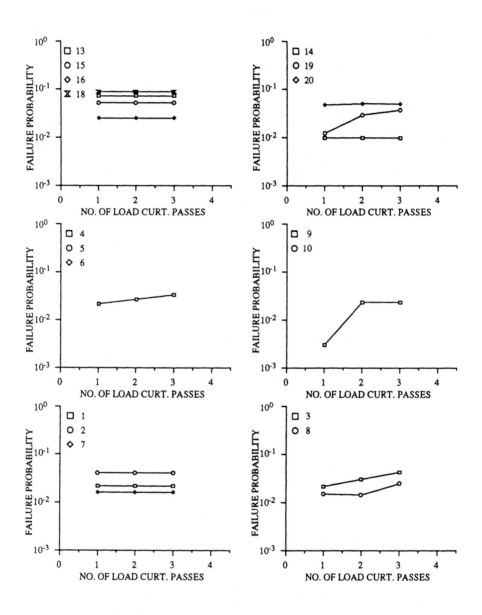

Figure 3.1 – Probability of failure vs. number of load
           curtailment passes for the RTS

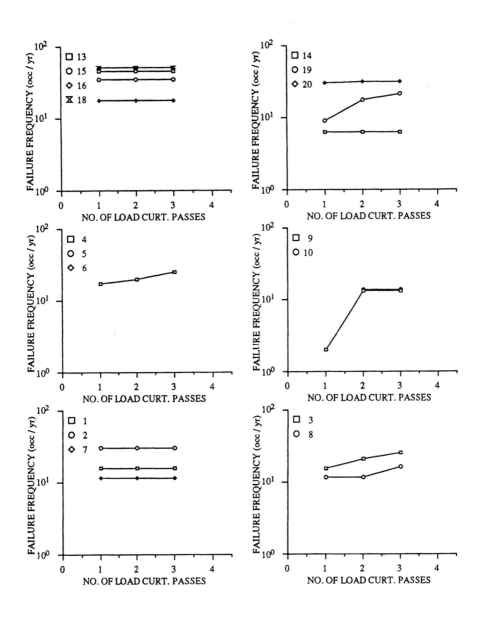

Figure 3.2 – Frequency of failure vs. number of load curtailment passes for the RTS

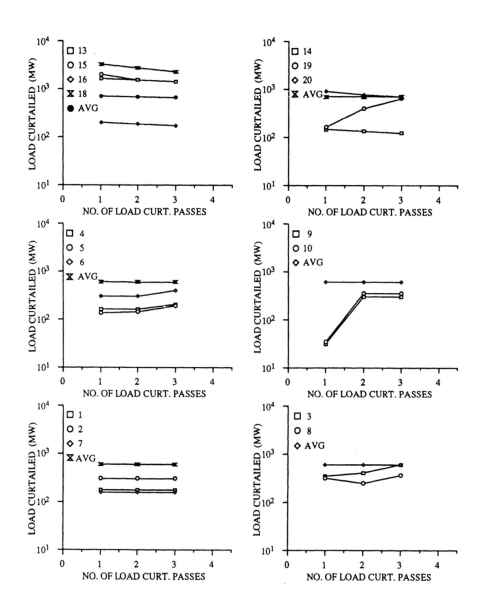

Figure 3.3 - Expected load curtailed in MW vs. number of
            load curtailment passes for the RTS

Table 3.1 - Expected load curtailed in MW for the RTS at various load curtailment passes

| Bus | Load Curtailment Pass | | | | |
| | 1st | 2nd | | 3rd | |
	Value	Value	Variation w.r.t. 1st pass	Value	Variation w.r.t. 1st pass
1	171.50	171.23	1.00	171.22	1.00
2	314.64	314.37	1.00	314.32	1.00
3	369.29	439.81	1.19	625.05	1.69
4	172.44	169.55	0.98	223.50	1.30
5	145.29	154.43	1.06	208.11	1.43
6	317.00	311.68	0.98	410.85	1.30
7	160.41	160.36	1.00	160.33	1.00
8	326.33	268.70	0.82	417.43	1.28
9	32.90	305.75	9.29	308.03	9.36
10	36.6	364.99	9.97	366.89	10.02
13	1769.22	1638.36	0.93	1534.73	0.87
14	150.49	138.60	0.92	127.89	0.85
15	1961.79	1662.60	0.85	1548.42	0.79
16	204.90	198.50	0.97	192.92	0.94
18	3377.93	2887.15	0.85	2481.43	0.73
19	166.44	396.46	2.38	543.14	3.26
20	853.40	652.54	0.76	558.44	0.65

ment of these buses whenever a generating unit is out either at bus 2 or at bus 7. However, at load curtailment pass 3, buses 4, 6 and 8 start sharing the load curtailment whenever a generating unit is removed from bus 13 or bus 23 with other generating units. The expected load curtailed at buses 13, 14, 15, 16, 18 and 20 further decreases because at pass 3, buses in the south and bus 19 share the load curtailment of these buses.

Appropriate Contingency Levels

A major concern in adequacy studies of a composite generation and transmission system is the selection and testing of outage contingencies which occur frequently and have a severe impact on the system performance [48]. In many cases, severity associated with a contingency event is inversely related to the frequency and the probability of its occurrence. In other words, as the number

of components involved in a simultaneous outage increases, both the probability and the frequency of the contingency decrease.   In the contingency enumeration approach,
a question often raised is whether the analysis is thorough enough and that a sufficient number of outage events
have been considered.   The selection of an 'appropriate'
outage level is therefore of fundamental importance in
the adequacy evaluation of a composite power system.   The
major objection to considering a large number of outage
events is the computation time required to solve these
contingencies.   This section examines contingency cutoff
criteria and the effect of considering higher level outage contingencies on the adequacy of the RTS.

Selection of an appropriate outage level is dictated
by various factors such as the size of the system, the
probabilities and the frequencies of the outage events,
the severity associated with an outage event, the purpose
of the adequacy studies, the computation time required to
evaluate each outage contingency, and the criteria used
to determine the system status, i.e. a failure or success
state.   The computation time increases rapidly as the
contingency level increases, particularly when an a.c.
load flow is used to analyze each contingency.   In order
to limit the number of contingencies, fixed criteria such
as the selection of single or double level contingencies
and/or variable criteria such as a frequency/probability
cutoff limit and/or a ranking cutoff, etc., are used.
Studies have shown that restricting contingencies to the
second level is not adequate particularly for a large
system, and that higher level outages should be considered.   Table 3.2 shows the sum of the probability of all
independent outage contingencies up to the 2nd level for
the RTS.   The sum of the probabilities for all possible
outage contingencies is always unity.   As shown in Table
3.2, the sum of the probabilities in the case of line
outages is close to 1.0, but the sum of the probabili-

Table 3.2 — Sum of the probabilities of all contingencies
         up to the 2nd level

Contingency description	Sum
Lines only	0.991130
Generators only	0.841244
Both lines & generators	0.834817

ties in the case of generator outages is considerably
less than unity. It can therefore be reasonably deduced
that as the size of a system increases, the calculation
of adequacy indices involving 1st and 2nd level contingencies, particularly for generator outages, will provide
optimistic results. This is due to the fact that, as the
number of generating units in a system increases, the
probability and the frequency of an independent outage
event involving three or more components increase to the
point at which these events cannot be ignored. The testing of higher level independent generator outages is,
therefore, necessary when calculating adequacy indices.
In this chapter, independent outages for generating units
up to the 4th level and independent outages for transmission lines and transformers up to the 2nd level are considered. The reasons for not considering higher outage
levels for transmission lines are as follows:

(a) as shown in Table 3.2, the sum of the probabilities
    for all transmission line contingencies up to the 2nd
    level is very close to unity,

(b) due to the system topology, transmission lines are
    also subjected to common cause failures such as the
    failure of a transmission tower supporting two or
    more transmission circuits or two or more transmis-
    sion circuits on a common right of way etc. Trans-
    mission lines are also exposed to adverse climatic
    conditions which create high failure rates. The
    effect of 'failure bunching' [52,54] due to adverse
    weather depends upon the network configuration and
    the meteorological conditions of the region. Outages

of transmission lines also result from station related failures. The contribution of these dependent
outages can be quite significant and many times larger than those of higher level independent transmission outages. It is therefore not practical or even
valid to consider the contribution of higher level
independent outages for transmission circuits and
ignore the contributions of more significant dependent outages. Modeling of dependent outages is
discussed in Appendix 3.

It is also possible to utilize a probability or frequency cutoff criterion, in addition to limiting the number of contingencies on the basis of outage level, i.e.,
those contingencies which have a frequency of occurrence
less than 1×10^{-9} are not solved as their contribution is
assumed to be negligible. The inclusion of higher level
generating unit outages can increase the CPU time considerably. This is shown in Table 3.3. One effective way
to reduce the computation time is to sort the identical
units and calculate the adequacy indices by solving the
a.c. load flow, if required, for only one contingency.
The contribution of the remaining identical contingencies
is calculated by multiplying the adequacy indices for
this contingency by the number of identical contingencies. Identical generating units are considered to have
the same MW rating, failure and repair rates and be at
the same generating station.

Table 3.3 – CPU time in minutes for the generator outages
            using a VAX–11/780 digital computer

Contingency level	Without sorting (minutes)	With sorting (minutes)
Up to the 1st level	0.14	0.12 (86%)
Up to the 2nd level	1.42	0.45 (32%)
Up to the 3rd level	16.80	2.27 (14%)
Up to the 4th level	137.14	14.14 (11%)

In order to realize maximum saving in CPU time, the identical generating units are arranged in a way such that contingencies involving identical units are evaluated before the contingencies involving non-identical generating units. This is easily done by placing identical units at the beginning of the evaluation list in terms of input to the digital computer program. Table 3.3 also shows the CPU time (in minutes) required for the generator outages, if identical units are sorted. A percentile comparison between the two sets of CPU time, with and without sorting identical units, is shown for each case. The quantities inside the brackets are percentage values of CPU time as compared to the CPU time shown in column 2. The saving in the CPU time due to sorting the identical units is quite significant. The sorting of identical units becomes even more significant as the depth of the contingency level increases and becomes a very effective way of reducing the CPU time. A further reduction in CPU time can be achieved by replacing n non-identical units connected at one bus and having equal MW rating but differing slightly in their failure and repair rates by n identical units, with each unit having the worst failure rate and the repair rate of the n units. This, however, provides pessimistic system adequacy indices.

The digital computer program developed at the University of Saskatchewan (COMREL) can consider simultaneous independent outages up to four generating units and two lines/transformers. The terms line and element outages are used interchangeably in this chapter and are intended to include both line and transformer outages. As with the program (RELACS) developed at UMIST, the level is user specified although the 4th level is generally used. Selection of this level is dictated, primarily, by the tremendous increase in computation time for higher levels and the marginal contribution of these outages to the adequacy indices. Table 3.4 shows the sum of the probabil-

ities for generator outages at different contingency
levels.

Table 3.4 – Sum of the probabilities for the generator
            outages

Contingency Level	Sum
1st	0.589915
2nd	0.841244
3rd	0.953814
4th	0.989544

The sum of the probabilities associated with contingencies up to the 4th level are 98.95% for the RTS. The
remaining 1.05% is contributed by contingencies beyond
the 4th level. The contribution of these higher level
outages can be calculated by solving an enormously high
number (201,376) of contingencies at the expense of ex-
cessive CPU time. In order to account for these higher
level contingencies (sometimes termed as more-off states)
the probability and the frequency of outage events at the
4th level can be modified such that they include the effect of successive states. A more-off state at a given
contingency level is a state in which at least one more
component is out of service in addition to those already
out at that level, e.g. for 2nd level independent outages, states representing the outage of three or more
components are designated as more-off states. In the
case of line outages, the probability and the frequency
at the 2nd outage level for each line contingency are
modified to take into account the contribution of the
more-off states.

The outage contingencies at the different levels considered in this chapter are designated as follows:
(a) 1st level outage contingency: outage of one system
    component such as a generating unit or transmission
    line,
(b) 2nd level outage contingency: 1st level outage con-

tingencies, simultaneous independent outages of any
two components, i.e. outages of two generating units,
outages of two transmission lines, and outages of one
generating unit and one line,

(c) 3rd level outage contingencies: 2nd level outage con-
    tingencies and simultaneous independent outages of
    three generating units,

(d) 4th level outage contingencies: 3rd level outage con-
    tingencies and simultaneous independent outages of
    four generating units.

Figures 3.4 and 3.5 show the variation in the two indices, frequency of failure and expected load curtailed
in terms of contingency level.

No bus in the RTS experiences a problem if only one
component is out of service. Even if two components are
out, buses 1, 2, 3 and 7 do not experience any load interruption as seen from Figure 3.5. The remaining buses,
however, experience load curtailment when two components
are out of service. Buses 4, 5 and 6 experience total
load curtailment whenever both the lines terminated at
these buses are out of operation. Since the probability
and the frequency of two lines being out of service are
quite small, the values of the expected load curtailed
are also small.

Buses 1 and 7 do not experience any load curtailment
due to the outages of two generating units anywhere in
the system. However, buses in the north (230 kV region)
experience load interruption when two large generating
units are removed from the system. Since all the large
generating units are concentrated in the north region and
because of the load curtailment philosophy, buses 13, 14,
15, 16, 18, 19 and 20 experience load curtailment. The
amount of load curtailment is proportional to the load
connected at each bus. Table 3.5 gives the number of load
curtailments and the expected load curtailed in MW at
each bus. The values of the expected load curtailed are

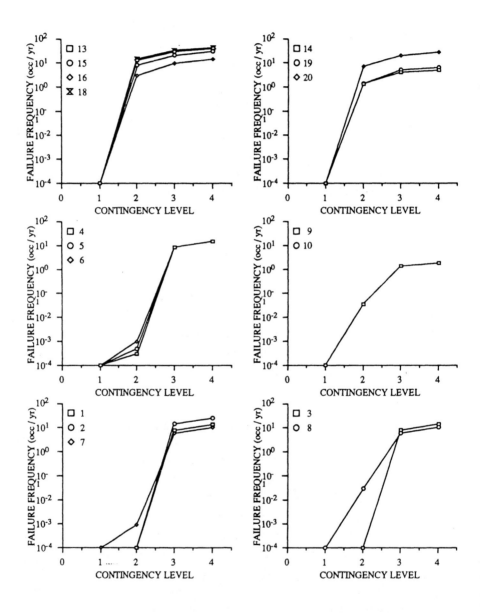

Figure 3.4 – Frequency of failure vs. contingency level
            for the RTS

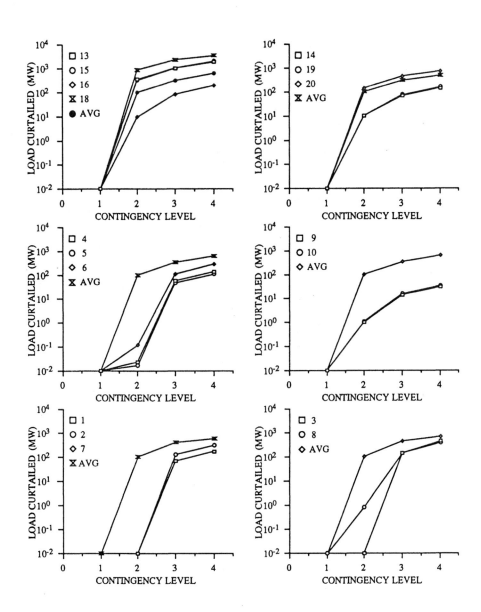

Figure 3.5 — Expected load curtailed in MW vs. contin-
        gency level for the RTS

Table 3.5 - Annualized bus indices for the RTS at various
            contingency levels

Bus	1st Cont. actual value	2nd Cont. actual value	3rd cont. Actual value	3rd cont. Incr. w.r.t. 2nd cont.	4th cont. Actual value	4th cont. Incr. w.r.t. 2nd cont.
		Number	of load	curtailments		
1	0.0000	0.0000	8.2191		16.5930	
2	0.0000	0.0000	16.3514		30.0112	
3	0.0000	0.0000	8.2191		16.7308	
4	0.0000	0.0004	8.2194	20548.5	16.5443	41360.8
5	0.0000	0.0005	8.2195	16439.2	16.5935	33187.2
6	0.0000	0.0010	8.2201	8220.1	16.5449	16544.9
7	0.0000	0.0000	6.1201		11.9833	
8	0.0000	0.0340	6.1540	181.0	12.0172	353.5
9	0.0000	0.0340	1.5422	45.4	1.9868	58.4
10	0.0000	0.0340	1.5422	45.4	1.9868	58.4
13	0.0000	12.7885	33.3278	2.6	45.8370	3.6
14	0.0000	1.3643	4.3085	3.2	6.7082	4.9
15	0.0000	8.7939	24.1961	2.7	35.3866	4.0
16	0.0000	2.8818	10.9981	3.8	18.3530	6.4
18	0.0000	15.1452	38.2298	2.5	51.5144	3.4
19	0.0000	1.3639	5.2735	3.9	8.0516	5.9
20	0.0000	7.5523	21.6245	2.9	29.9752	4.0
		Total	Load	Curtailed (MW)		
1	0.0000	0.0000	69.2099		171.5099	
2	0.0000	0.0000	130.1699		314.6499	
3	0.0000	0.0000	135.4799		369.2900	
4	0.0000	0.0300	65.0000	2166.7	172.4400	5748.0
5	0.0000	0.0200	53.4599	2673.0	145.2899	7264.5
6	0.0000	0.1400	119.5500	853.9	317.0000	2264.3
7	0.0000	0.0000	75.6900		160.4100	
8	0.0000	0.8300	137.1100	165.2	326.3299	393.2
9	0.0000	0.8500	15.4099	18.1	32.9000	38.7
10	0.0000	0.9500	17.1800	18.1	36.6599	38.6
13	0.0000	349.8500	1145.0300	3.3	1769.2199	5.1
14	0.0000	10.0699	74.0299	7.4	150.4900	14.9
15	0.0000	316.9200	1184.2600	3.7	1961.7900	6.2
16	0.0000	12.3199	100.6399	8.2	204.8999	16.6
18	0.0000	830.9699	2337.3400	2.8	3377.9299	4.1
19	0.0000	9.3199	80.3000	8.6	166.4400	17.9
20	0.0000	137.7500	550.2999	4.0	853.4099	6.2

Incr. = Increment   Cont. = Contingency

many times higher for buses in the north as compared to
those in the south. Buses in the north experience load
curtailment in the event of two generating unit outages
while buses in the south only experience load curtailment
when either three generating units are out or at least
one line in combination with other component(s) is out.
The load curtailment at buses 8, 9 and 10 for the 2nd
outage level is due to the outage of line 11 in combination with the outage of any large (> 350 MW) generating
unit in the system. Whenever three generating units
involving at least one unit from the south region are out
of operation, buses in the south also encounter load interruption. At the same time, the amount of load curtailed at buses in the north also increases by at least
200% over its value at the 2nd outage contingency level.

The increment in the adequacy indices for buses in
the south is quite high as seen from Table 3.5. This
non-uniform trend in the variation of the adequacy indices also continues at the 4th outage level. It is quite
clear from a study of Figures 3.4 and 3.5 and Table 3.6
that for this system, the calculation of the 4th level
outage contingencies is necessary for both the bus indices and the system indices. Table 3.6 gives some of the
system indices and the corresponding increment with respect to the 2nd outage level contingencies for this system. As observed from Table 3.6, the value of the severity index at the 4th outage level is about four times
higher than the value at the 2nd outage level, while the
value of the average load curtailed for each load point
per year is more than six times the value at the 2nd outage level. This is due to the fact that the number of
load curtailment contingencies increases tremendously as
the outage level increases. This is shown in Table 3.7,
which gives the actual number of contingencies for different outage levels.

The average values of the bus indices at different

Table 3.6 — System indices for the RTS at various contin-
            gency levels

1st	2nd	3rd cont.		4th cont.	
Cont. actual value	Cont. actual value	Actual value	Incr. w.r.t. 2nd cont.	Actual value	Incr. w.r.t. 2nd cont.

Bulk power supply disturbances

| 0.00000 | 17.30045 | 46.02694 | 2.7 | 62.80533 | 3.6 |

Total probability

| 0.60023 | 0.83869 | 0.94994 | 1.1 | 0.98223 | 1.2 |

IEEE indices

Bulk power interruption index (MW/MW-yr)

| 0.00000 | 0.58597 | 2.20707 | 3.8 | 3.69497 | 6.3 |

Bulk power energy curtailment index (MWh/yr)

| 0.00000 | 12.34820 | 35.30506 | 2.9 | 51.16785 | 4.1 |

Bulk power supply average MW curtailment index
(MW/disturbance)

| 0.00000 | 96.53047 | 136.66216 | 1.4 | 167.67130 | 1.7 |

Modified bulk power energy curtailment index

| 0.00000 | 0.00141 | 0.00403 | 2.9 | 0.00584 | 4.1 |

Severity index (system-minutes)

| 0.00000 | 740.89203 | 2118.30396 | 2.9 | 3070.07105 | 4.1 |

Average indices

Av. no. of hrs of load curtailment/load pt./yr

| 0.00000 | 54.90038 | 181.01126 | 3.3 | 255.69083 | 4.7 |

Av. no. of load curtailments/load pt./yr

| 0.00000 | 2.94081 | 12.39800 | 4.2 | 19.81285 | 6.7 |

Av. load curtailed/load pt./yr-MW

| 0.00000 | 98.23650 | 370.00827 | 3.8 | 619.45007 | 6.3 |

Av. energy curtailed/load pt./yr-MWh

| 0.00000 | 2070.13940 | 5918.79004 | 2.9 | 8578.14063 | 4.1 |

Incr. = Increment   Cont. = Contingency

Table 3.7 — Number of contingencies for the RTS at various contingency levels

Description	Contingency level			
	1st	2nd	3rd	4th
No. of generator contingencies	32	528	2488	41448
No. of line contingencies	38	741	741	741
No. of G—L contingencies considered	0	1178	1178	1178
No. of voltage violation contingencies	0	0	0	0
No. of MVAR limit violation contingencies	1	79	586	6281
No. of no—convergence contingencies	0	1	1	1
No. of load curtailment contingencies	0	28	787	11996
No. of bus isolation contingencies	1	75	75	75
No. of split network contingencies	0	1	1	1
No. of firm load curtailment contingencies	0	8	166	3240

contingency levels are mainly influenced by the indices of the buses in the north (230 kV) region. The marginal increment of the average values of the adequacy indices may, therefore, give some idea about the behavior of the bus indices for buses in the north region but drawing any conclusion about the indices for buses in the south from the average values is highly misleading as seen from Figures 3.5 and 3.6. This is also the case with the system indices. Approximately 80% of the contribution to the system indices comes from buses in the northern area. This reinforces the observation made earlier that it may be erroneous to draw conclusions about the bus indices from either system indices or average values of the bus indices.

Station Originated And Common Cause Events

Background. Recent attention to the role of terminal stations in system disturbances strongly supports the

need to recognize the multiple outages of major system
components due to terminal station failures and common
cause outages in composite system adequacy assessment.
Landgren and Anderson reported [55] that over 40% of the
multiple line outages in the Commonwealth Edison Company's 345 kV power system are caused by terminal related
disturbances. The probabilities associated with station
originated multiple line outages can be quite high compared to the corresponding event probabilities associated
with independent outages. A similar effect occurs with
common cause outages. It is therefore not practical to
consider higher level independent line outages and ignore
the station originated multiple outages and common cause
outages. The effect of stations is sufficiently dominant
in most cases that their inclusion diminishes the need to
consider high level independent line outages. A detailed
discussion of station related outages and common cause
outages is given in Appendix 3.

   Station Originated Outages. Present composite system
reliability evaluation techniques normally include terminal stations by simply modifying the reliability parameters of lines and/or generators affected by the outages
of station components. This approach is suggested in the
original reference describing the RTS [58]. This is correct only if the failure of a station component results
in the outage of a single major element. When the unavailability of two or more elements in a system is considered, the approach assumes an unrealistic independence
between these element outages. A more realistic approach
is to treat the major component resulting from terminal
station failures as separate events. It is not feasible
to solve all the contingencies simultaneously due to station initiated outages and independent outages. System
event data can be obtained due to failures originating in
the terminal stations. These event data include:

(a) various sets of major components on outage due to station related failures,
(b) probability of outage of each set of major elements,
(c) frequency of outage of each set of major elements.

The following assumptions were utilized in the development of the station originated models [43] used for the studies illustrated in this section:

(a) the probability of a stuck breaker condition is sufficiently small that it can be assumed to be zero,
(b) the effects of adverse weather are neglected,
(c) the probability of overlapping outages of three or more station components is negligible and is therefore not considered,
(d) a component is not taken out for preventive maintenance if it results in the outage of a major component,
(e) relays associated with all breakers are assumed to be non directional.

The failure of a station component may result in multiple outages of generators and/or transmission lines. It may also isolate the load feeder completely. Several station related failure events could result in the same system effect i.e. they might cause the same set of major elements or load feeders to be removed from the system. In this case, the reliability indices (probabilities and frequencies of outage) associated with these failure events can be directly added for the purposes of composite system adequacy evaluation.

The station originated models were developed [43] for the following failure events originating in the stations:

(a) forced outage (active and passive failure) of a breaker,
(b) forced outage of a bus section,
(c) forced outage of a transformer,
(d) overlapping forced outages of breakers, transformers and bus sections,

(e) forced outage of a component overlapping maintenance
    of another component.

A general list of the major component combinations
removed from service due to station disturbances is as
follows:

(a) one generating unit is out,
(b) two generating units are out,
(c) three generating units are out,
(d) four generating units are out,
(e) one line or transformer is out,
(f) two line(s) or transformer(s) are out,
(g) three line(s) or transformer(s) are out,
(h) one generating unit and one line or transformer is
    out,
(i) two generating units and one line or transformer is
    out,
(j) load feeder(s) are isolated.

The contribution of the above contingencies, except
for the outage of three lines (No. g), the outage of two
generating units and one line (No. i), and the isolation
of the load feeder (No. j) is taken into account by adding the probabilities and the frequencies to the probability and the frequency of the respective contingencies
resulting from the independent outages of these components. The remaining three outages are calculated separately. The effect of station originated outages on the
adequacy indices of the RTS is illustrated below.

The original single line diagram of the RTS is shown
in Appendix 1. The original network has been extended in
Appendix 2 which shows the complete network with station
configurations at each bus. These station configurations
have been adapted from existing configurations in actual
systems.

System studies were conducted at load levels of 2400
MW, 2550 MW, 2700 MW and 2850 MW in order to examine the
effect of terminal stations at different load levels. A

description of the load model for a peak load of 2850 MW
is given in Appendix 1. The load model at other peak
loads was obtained by scaling down the basic load data.

Two basic annualized individual load point indices,
i.e. the total number of load curtailments and the expected load curtailed in MW, are given in Tables 3.8 and
3.9, respectively for the system load of 2400 MW.

The indices shown in column 2 (without stations) in
Tables 3.8 and 3.9 are due to the independent outages of
the generating units up to the 4th level and transmission
lines and transformers up to the 2nd level. The indices
noted in column 3 (with stations) were obtained after including the effect of station originated outages together
with the independent outages of the generating units,
lines and transformers. The net increase in the indices
is given in column 4 of each table. As seen from the
tables, the effect of station originated outages on the
bus indices is considerable. Buses 9 and 10 which do not
encounter load curtailment problems with independent outages, experience load curtailment because of the failure
of station components. It can be seen that the effect of
station originated outages on the indices at each bus is
not uniform. The effect depends upon the load curtailment philosophy for an outage event and the station configuration selected at a bus.

  Common Cause Outages. The circuits considered to be
vulnerable to common cause outages are as follows:
(a) lines 12 and 13,
(b) lines 18 and 20,
(c) lines 25 and 26,
(d) lines 31 and 38,
(e) lines 32 and 33,
(f) lines 34 and 35,
(g) lines 36 and 37.
  Table 3.10 shows system indices for the four cases:

Table 3.8 – Expected annualized number of load curtail-
          ments at the load buses in the RTS

	System load = 2400 MW		
Bus	Without stations	With stations	% increase
2	0.669	0.705	5.38
3	0.334	0.370	9.73
4	0.335	0.433	29.25
5	0.335	0.433	29.25
6	0.336	0.434	29.16
7	0.355	0.363	2.25
8	0.355	0.400	12.67
9	0.0	0.0367	I
10	0.0	0.0369	I
13	1.997	2.033	1.80
14	0.233	0.284	21.88
15	1.962	2.000	1.93
16	0.538	0.587	9.10
18	2.216	2.256	1.80
19	0.232	2.256	1.80
20	1.162	1.257	8.17

Table 3.9 – Expected annualized load curtailed at the
          load buses in the RTS

	System load = 2400 MW		
Load bus no.	Without stations	With stations	% increase
1	1.62 MW	4.77 MW	194.4
2	3.22	6.23	93.47
3	2.70	8.06	198.52
4	1.24	7.12	474.2
5	1.10	7.00	536.37
6	2.34	13.19	463.67
7	2.64	3.46	31.06
8	4.16	10.65	156.00
9	0.0	5.33	I
10	0.0	5.91	I
13	41.86	52.23	24.77
14	2.64	11.09	320.07
15	50.31	63.16	25.54
16	3.43	7.65	123.03
18	67.15	81.18	20.89
19	2.34	10.97	368.80
20	19.17	29.98	56.39

Table 3.10 – System indices for the RTS with/without com-
            mon   cause   and   station   originated   outage
            events

			System load = 2400 MW			
Case (1) independent outages	(2) C.C. outages		(3) S.O. outages		(4) C.C. and S.O. outages	
Actual value	Actual value	Incr. w.r.t. (1)	Actual value	Incr. w.r.t. (1)	Actual value	Incr. w.r.t. (1)

IEEE indices

Bulk power interruption index (MW/MW-yr)

| 0.08579 | 0.10308 | 1.20 | 0.13666 | 1.59 | 0.15803 | 1.84 |

Bulk power energy curtailment index (MWh/yr)

| 1.18503 | 1.30798 | 1.10 | 2.05153 | 1.73 | 2.24482 | 1.89 |

Bulk power supply average MW curtailment index
(MW/disturbance)

| 92.81903 | 100.34611 | 1.08 | 104.54426 | 1.13 | 110.08784 | 1.19 |

Modified bulk power energy curtailment index

| 0.00014 | 0.00015 | 1.10 | 0.00023 | 1.73 | 0.00026 | 1.89 |

Severity index (system-minutes)

| 71.10200 | 78.47900 | 1.10 | 123.09200 | 1.73 | 134.68900 | 1.89 |

Average indices

Av. no. of hrs of load curtailment/load pt./yr

| 8.71335 | 9.14968 | 1.05 | 9.56775 | 1.10 | 10.19681 | 1.17 |

Av. no. of load curtailments/load pt./yr

| 0.67052 | 0.73267 | 1.09 | 0.72322 | 1.08 | 0.79649 | 1.19 |

Av. load curtailed/load pt./yr-MW

| 12.11181 | 14.55229 | 1.20 | 19.29249 | 1.59 | 22.31119 | 1.84 |

Av. energy curtailed/load pt./yr-MWh

| 167.29817 | 184.65901 | 1.10 | 289.62839 | 1.73 | 316.92191 | 1.89 |

Incr. = Increment   C.C. = Common cause
S.O. = Station originated

(a) independent outages only,
(b) independent and common cause,
(c) independent and station originated outages,
(d) independent,   common   cause   and   station   originated
    outages.

The increment in the system indices in each case with respect to case (1) is also given in Table 3.10. The effect of the station originated outage events is more pronounced as compared to the effect of common cause outage events as seen from Table 3.10. The major contribution of the station originated outages comes from the contingencies resulting either from the isolation of a load or from the removal of two or more large generating units because of a fault in the station components. The inclusion of common cause outages results in the following two split network situations:

(a) outage of lines 36, 37, and 29 curtails total load at
    bus 19 and 20,
(b) outage of lines 25, 26 and 28 creates capacity defi-
    ciency in the separated network having buses 1 to 16
    and buses 19, 20 and 23. Most of the buses in the
    network experience load curtailment.

The number of load curtailment contingencies increases from 365 to 414 when common cause outages are considered. The combined effect of common cause and station originated outages is quite large. The severity index is approximately 1.9 times its value for independent outage events.

The system indices after including the effect of station originated and common cause outages were also calculated at four system loads, 2850 MW, 2700 MW, 2550 MW and 2400 MW as shown in Table 3.11. Table 3.12 gives the system indices at these loads for independent outage events without including the effect of station originated outages and common cause outages. As seen from Tables 3.11 and 3.12, the absolute value of the indices with the

Table 3.11 – System indices for the RTS at various load
            levels including common cause and station
            originated outage events

System load in MW						
2850.0	2700.0		2550.0		2400.0	
Actual value	Actual value	Decr. w.r.t. (1)	Actual value	Decr. w.r.t. (1)	Actual value	Decr. w.r.t. (1)

IEEE indices

Bulk power interruption index (MW/MW-yr)
| 3.71753 | 1.47203 | 0.40 | 0.46006 | 0.12 | 0.13666 | 0.04 |

Bulk power energy curtailment index (MWh/yr)
| 51.57207 | 20.87945 | 0.40 | 6.32895 | 0.12 | 2.05153 | 0.04 |

Bulk power supply average MW curtailment index
(MW/disturbance)
| 168.41957 | 158.73817 | 0.94 | 58.05872 | 0.34 | 58.05872 | 0.34 |

Modified bulk power energy curtailment index
| 0.00589 | 0.00238 | 0.40 | 0.00072 | 0.12 | 0.00023 | 0.04 |

Severity index (system–minutes)
| 3094.32398 | 1252.76697 | 0.40 | 379.73700 | 0.12 | 123.09200 | 0.04 |

Average indices

Av. no. of hrs of load curtailment/load pt./yr
| 285.24514 | 101.67792 | 0.39 | 79.23013 | 0.31 | 9.56775 | 0.04 |

Av. no. of load curtailments/load pt./yr
| 19.95304 | 7.62601 | 0.38 | 6.11028 | 0.31 | 0.72322 | 0.04 |

Av. load curtailed/load pt./yr-MW
| 623.23340 | 233.79334 | 0.38 | 69.00830 | 0.11 | 19.29249 | 0.03 |

Av. energy curtailed/load pt./yr-MWh
| 8645.90723 | 3316.14795 | 0.38 | 949.34283 | 0.11 | 289.62839 | 0.3 |

Decr. = Decrement   C.C. = Common cause
S.O. = Station originated

Table 3.12 – System indices for the RTS at various loads
           levels without including common cause and
           station originated outage events

	System load in MW					
2850.0	2700.0		2550.0		2400.0	
Actual value	Actual value	Decr. w.r.t. (1)	Actual value	Decr. w.r.t. (1)	Actual value	Decr. w.r.t. (1)

IEEE indices

Bulk power interruption index (MW/MW-yr)

3.69497	1.43433	0.39	0.41150	0.11	0.08579	0.02

Bulk power energy curtailment index (MWh/yr)

51.16785	20.23907	0.40	5.50024	0.11	1.18503	0.02

Bulk power supply average MW curtailment index
(MW/disturbance)

167.67130	158.89359	0.95	53.80362	0.32	92.81903	0.55

Modified bulk power energy curtailment index

0.00584	0.00231	0.40	0.00062	0.11	0.00014	0.02

Severity index (system-minutes)

3070.07105	1214.34399	0.40	330.01401	0.11	71.10200	0.02

Average indices

Av. no. of hrs of load curtailment/load pt./yr

255.69083	101.86563	0.40	79.20758	0.31	8.71335	0.03

Av. no. of load curtailments/load pt./yr

19.81285	7.63679	0.39	6.11366	0.31	0.67052	0.03

Av. load curtailed/load pt./yr-MW

619.45007	277.80481	0.37	61.72573	0.10	12.11181	0.02

Av. energy curtailed/load pt./yr-MWh

8578.14062	3214.4409	0.37	825.0358	0.10	167.29817	0.02

Decr. = Decrement   C.C. = Common cause
S.O. = Station originated

station originated and common cause outages increases at each system load level. The decrement in the indices with decrease in the load when considering common cause and station originated outages is not as large as it is when only independent outages are considered. This is due to the fact that as the load decreases, the contribution to the adequacy indices comes mainly from the isolation of a load. Faults in the station components are largely responsible for the isolation of a load. However, load is also isolated when lines terminated at a load bus are removed. No system bus experiences total load curtailment for the outage of one line only. The number of bus isolation contingencies with the inclusion of common cause and station originated outage events has increased from 48 to 104.

The probability and the frequency of isolation of a bus due to a fault in the station components is higher than the values due to the removal of the associated transmission lines. Therefore, the contribution to the adequacy indices of the bus isolation contingencies when common cause and station originated outages are considered is quite significant.

ALTERNATIVE METHODS

Background

A wide range of techniques have been proposed for reliability evaluation in composite generation and transmission systems. Many of the proposed methods are basically variations on fundamental approaches to the problem. Over the years, however, it has become obvious that there are in general, basic conceptual differences between the techniques used in Europe, particularly Italy and France, and those utilized in North America. This section briefly illustrates some of the fundamental differences between these techniques for composite system reliability evaluation by application to the RTS [46].

The generation and load data and the transmission configuration are described in detail in Appendix 1.   Only
those data which are required to provide a better interpretation of the analysis conducted are presented in this
chapter.   The two techniques presented are conceptually
different.   The first one illustrates the Monte Carlo approach used by ENEL in Italy in its planning practice.
The second approach is the contingency evaluation procedure developed at the University of Saskatchewan.   The
two methods are representative of other evaluation methods found in Europe and North America respectively.   The
basic objective is to illustrate the comparative fundamental differences between the approaches and show the
numerical results obtained in each case using a representative test system.

The calculated results shown for the RTS obtained using the two methods are not directly comparable.   The basic differences are due to the load models used, the network representation and solution technique and the overload alleviation methods employed.   In the case of the
RTS, the load reduction philosophy utilized in the case
of line overload and inadequate installed capacity is a
major factor.   The calculated individual load point adequacy indices can vary widely with the load curtailment
policy employed.   The calculated indices are relative
values which respond to various factors that influence
the adequacy of the system and can be used to consider
planning alternatives and to assess the sensitivity of
the system to configuration changes.

Method 1 – The ENEL Approach

Concept Of Method.   The "cost of reliability" is one
of three components used by ENEL in computing the total
system cost.   The other two components used in system op-
timization are the capital and operating costs.   The cost
of reliability is obtained by developing a risk index in

the form of the expected yearly curtailed energy (kWhr/ yr). This is then transformed into a cost value using an appropriate conversion factor. A widely accepted value for this unit cost is about 1$/kWh. In regard to the composite system, the annual energy curtailed can be obtained:

(a) for the entire system − with respect to the causes i.e. static deficiency of generation, transmission or transformation,

(b) for each network node where a deficiency occurs,

(c) for the weather conditions during which a deficiency occurs.

The basic technique used to accomplish this is the Monte Carlo method. The main advantage of this method is the feasibility of taking into account theoretically any random variable, any contingency and of adopting operation policies similar to the real ones. The disadvantage could be the required computing time. It also allows a better "dialogue" within the source utility between planning and operation departments, since the factors/indices used are nearly the same. A specified calendar year of system life is examined using repeated "yearly samples" each consisting of 8760 hours which are randomly selected. The computer program (SICRET) has been extensively described in the literature. It has been used for approximately 10 years [8,9,12,13,17] and is currently utilized in network planning of ENEL's system.

The models used for system components and dispatching actions are as follows:

(a) load model − the real power load for the system and for each load point is determined for each hour. A further subdivision is made at each bus indicating the percentage of load that can be interrupted. No reactive load is considered in the analysis,

(b) environment model − three weather conditions can be considered in each homogeneous area of the system.

These three conditions designated as fair, bad and
stormy weather are considered to randomly alternate
during the annual study period,

(c) component failure model - forced outages of genera-
    tors, transformers, and other network components are
    represented by two state (up-down) models. The
    weather conditions are incorporated in the transmis-
    sion line models,

(d) generating unit model - units are characterized by
    their ratings, technical minimum loadings and hourly
    production costs,

(e) operating policy models - the operating policies must
    be similar to those performed in practise but must be
    sufficiently simple to permit the results to be ob-
    tained within reasonable computing times. The fol-
    lowing two types, selected by input, can be utilized:

        (i) pure safety policy,
       (ii) mixed economy - safety policy.

The two types differ in the policies adopted for unit
commitment and generation dispatch to relieve component
overload. Type (1) is utilized when only system reliability is required. Type (2) is utilized when the system operating costs are also required. Of course the
corresponding computing times are different.

Program Procedure. The basic steps in the program
are as follows:

(a) the system load is matched to the total available
    generation, by curtailing if necessary a portion of
    the load. This provides an overall risk index due to
    "lack of real power",

(b) if a mixed "economy-safety" policy is selected, the
    generating units are loaded in accordance with a
    priority list based on operating costs. Local con-
    straints such as limited energy availability can be
    included. If the "pure-safety" policy is selected,

the available units are loaded in proportion to their available capacity.

(c) the system network solution is obtained using a dc load flow. Consequently, the SICRET program is particularly suited to large meshed systems with no voltage/stability problems, as is the case for many systems of Western Europe. Cautions are to be taken in other cases,

(d) if the load on a component exceeds its capability, the generation dispatch is modified in an attempt to alleviate the overload. The overload relief is carried out according to the operation policy chosen (mixed economy-safety or pure safety),

(e) component overloads above an "alarm level" are tolerated up to a preselected capability limit. If this level is exceeded, certain loads are curtailed until the component loads come within this limit (this gives rise to a risk index due to "component overload"),

(f) the resulting energy curtailments in kWh/yr are summed for all the 8760 hours examined in the yearly sample. An estimate of the "average annual energy curtailed" is obtained by repeating the analysis for many yearly samples of the same solar year,

(g) when applying the mixed "economy-safety "policy and attempting to alleviate component overloads, capacity shifting is performed in such a way that the cost of shifting is a minimum. If the "pure-safety" policy is utilized, generation shifting (re-dispatch) is carried out only to produce maximum influence on the overload components with no attention to the system operating cost.

Application To RTS. When the transmission network is assumed to be completely reliable and capable, the overall system characteristics are shown in Table 3.13.

Table 3.13 - Generation system adequacy indices

Annual Peak	= 2,850 MW (at the load points)
Annual Energy	= 15,370 GWh/yr
Installed Capacity	= 3,405 MW
Capacity Reserve	= 19.5%
Expected Curtailed Energy	= 4,000 MWh/yr

The analysis of overall system reliability has been performed utilizing a dc load flow. The static adequacy of the system is evaluated assuming that the system does not present any voltage problems. The cost figures presented below were obtained using the mixed "economy-safety" policy for generation dispatch and overload relief. Table 3.14 shows the annual production and running cost of each type of generating unit and for the system(+). The overall annual energy curtailed is also shown, subdivided in accordance with the causes which produced the curtailments.

(+) Fuel prices at Italian costs at 1.1.1983. Rate of exchange: 1US$ = 1400 lira.

Table 3.14 - Annual production costs

Unit type	Capacity MW	Fuel cost mill/kWh	Energy GWh/yr	10^6 Cost lira/yr
Nuclear	800	8	5,442	43.536
Coal	1274	22.7	8,242	187.093
Oil	951	40.0	1,646	65.840
Hydro	300	–	25	–
Combustion turbine	80	95.7	14	1.340
Yearly curtailed energy				
Lack of generation			4.1	
Lack of capability			0.2	
		Sub Total	4.3	
Total	3,405		15,373	297.809

Table 3.15 shows the yearly curtailed energy subdivided according to the causes and where the curtailments occurred.

The following comments can be made regarding the re-

Table 3.15 – System curtailments

Lack of generation	Curtailment
Bus	MWh/yr
18	2,782
16	560
15	270
13	210
7	200
1	65
2	55
8	–
Sub Total	4,142

Overload	Curtailment
Bus	MWh/yr
5	66
4	50
19	–
20	–
8	100
7	–
Sub Total	216

Total energy curtailment = 4,358 MWh/yr

sults and the RTS based upon the analysis conducted:

(a) a great unbalance exists between the curtailments due
to the lack of generating capacity and those due to
overloads,

(b) the critical buses are those with load and only one
unit installed. In the case of forced outage, unbalance results between generation and load and the ne-
cessity of load shedding occurs. This is the case at
buses 18 and 16. At bus 13, where three units of 197
MW each are installed, a reasonable generation margin
(48%) exists even with one of the three units out of
service. The likelihood of load shedding in this case
is much lower,

(c) the transmission system in the RTS appears to be

somewhat oversized for the generation capacity involved. Curtailments only occur at buses 8, 4 and 5.
If a critical generation situation occurs on the
overall network which requires maximum output for the
units at bus 7 (300 MW), this entails overload of the
radial line between buses 7 and 8.

A logical extension of this study, in the manner normally conducted by ENEL, would be to utilize the program
to examine system modifications. The results shown in
Tables 3.14 and 3.15 indicate that considerable load
shedding occurs due to the lack of generation. The system generation composition has been tentatively modified
by adding a combustion turbine of 20 MW at bus 8. In
this case, the annual curtailed energy due to lack of
generation drops to 3.7 MWh for a total of 3.9 MWh/yr:
the main improvements are obtained in buses 18, 16, 8.
Additional studies can be conducted using the program.
The Monte Carlo approach is well suited to the determination of expected production costs and energy curtailments
occurring in a reference system and under conditions of
generation and transmission modifications. The incorporation of curtailed energy costs can be used to examine
potential alternatives with the intention of developing
an "optimum" configuration. The program is one in a family of programs used by ENEL in system planning. The
conducted studies would be used in conjunction with those
of load flow, stability, short circuit etc. to arrive at
investment decisions.

Method 2 - The U Of S Approach

Concept of Method. The primary objective in this approach is to evaluate adequacy indices for the total system and at every load point in the system [24]. Failure
of service at any busbar in the system includes the violation of the minimum acceptable voltage at the busbar
and/or failure of the system to supply the total load

connected to that busbar after alleviating the line overloads, generator MVAr limit violations etc.   Inadequacy
of the available generation to meet system load requirements is also considered as a failure at each busbar.
The evaluation of a composite system involves the analysis of all possible contingency states.   This of course
is not possible and therefore the number of contingencies
considered must be limited.   In addition to considering
the independent events associated with the outage of generating units, lines and transformers either singly or in
combination with others, common mode outages of multiple
elements [30] or multiple outages due to station originated events [32] can be included in the list of contingencies studied.

The generating schedules at appropriate generating
buses are modified for each generating unit outage case
to compensate for the loss of generation.   The violation
of a specified service quality criterion are determined
for each outage condition by conducting an ac load flow
analysis.   The bulk of the computational time is utilized
in the load flow analysis of each outage condition.   The
fast decoupled load flow technique is utilized in this
approach.   One problem encountered in composite system
adequacy evaluation is the overload of transmission lines
and transformers.   The simplest solution to this problem
is to allow these elements to operate in the overloaded
condition.   This results in optimistic reliability indices particularly when the line overload is heavy and per-
sists for a long duration.   An alternative solution is to
remove the overloaded component and continue to analyze
the remaining system until no other element is overloaded, or until the total system fails.   This approach results in pessimistic reliability indices which may suggest expensive and unnecessary investment in system improvement.   The approach used in this regard is to alleviate the line overloads by (a) generation rescheduling

and/or (b) curtailment of some of the loads.

When load curtailment is required, the numerical indices calculated at each bus can vary widely when different load curtailment philosophies are used [27]. This
makes it difficult to compare the results of one computer
program with those of another.

The procedure used in COMREL (U of S program) to cur-
tail load is, as described earlier in this chapter, a
flexible, pre-specified approach. The load at each bus
is classified into two types, firm and curtailable. In
the case of a deficiency in generation capacity, the curtailable load is interrupted first followed if necessary
by curtailment of firm load. The regional extent to which
a particular capacity deficiency is applied can be varied
using the three topological classifications designated as
load curtailment pass 1, 2, or 3. The expected load curtailed at the individual bases in the network can therefore change quite considerably with the prespecified mix
of firm/curtailable load and the regional constraints in
load curtailment.

Application To RTS. The system adequacy indices in
terms of individual bus values and overall system indices
are shown in Tables 3.16 and 3.17 respectively.

The full range of indices described in References 24
and 54 can also be obtained.

Independent overlapping outages up to the fourth level for generating units and up to the second level in
transmission elements were considered. The effects of
higher order generating unit failures were included by
using a cumulative probability and frequency for the
highest level contingency considered. Common mode or
common cause outages were not included in these studies.
The indices have been evaluated considering a single step
load model. This load model provides a relative estimate
of inadequacy at each load point and for the total system

Table 3.16 - Annualized bus indices for the RTS

System load - 2850 MW

Bus no.	Failure probability	Failure frequency occ/yr	Load curtailed MW	Energy curtailed MWh
1	0.022446	16.59	171.5	2,086
2	0.040999	30.01	314.6	3,827
3	0.022640	16.73	369.2	4,560
4	0.022394	16.54	172.4	2,133
5	0.022446	16.54	145.2	1,794
6	0.022395	16.54	317.0	3,920
7	0.015922	11.98	160.4	1,905
8	0.015950	12.01	326.3	3,972
9	0.003171	1.98	32.4	425
10	0.003171	1.98	36.6	474
13	0.071273	45.83	1,769.2	23,662
14	0.009556	6.70	150.4	1,793
15	0.056509	35.38	1,961.7	28,069
16	0.026011	18.35	204.9	2,478
18	0.083433	51.51	3,377.9	50,912
19	0.011667	8.05	166.4	2,017
20	0.046213	29.97	853.4	11,792
				145,819

Table 3.17 - Annualized system indices

Index description	Index value
IEEE Indices	
Bulk power interruption index (MW/MW-yr)	3.69
Bulk power energy curtailment index (MWh/yr)	51.16
Bulk power supply average MW curtailment index (MW/disturbance)	167.67
Energy index of unreliability	0.005841
Severity index (system-minutes)	3070.1

and provides representative indices to assess the sensitivity of the system adequacy to configuration changes. The adequacy indices are very sensitive to the load level. As an example, the Severity Index drops to 71.1 system-minutes at a load level of 2400 MW. A bus failure under any outage condition is defined as including the

violation of the acceptable voltage limits at that bus
and/or not meeting the load requirements at that bus,
generator MVAr limit violations and non-convergent situations etc. The swing bus overload, if any, is alleviated
by curtailing the load at various load buses. In these
studies, curtailable load at each bus is assumed to be
20% of the total load of the bus and the number of load
curtailment passes is specified as one. Line or transformer overload conditions are alleviated by generation
rescheduling and/or load curtailment at the buses. Bus 6
experiences load curtailment because of the overloading
of the connection between bus 6 and bus 10 when the line
between bus 6 and bus 2 is out. The outage of a 400 MW
generator at bus 18 results in unsatisfactory system performance because of generator MVAr limits. The effect of
line outages except that of outage of line 5 on bus indices is negligible to moderate. Buses 6, 13, 15, 18 and
20 have high inadequacy indices compared to the indices
of the other buses. Buses 3 and 6 also experience voltage violations. The maximum and minimum limits of bus
voltages were assumed to be 1.05 p.u. and 0.95 p.u. respectively.

The outage of the largest generating unit (400 MW)
alone does not cause load curtailment but the outage of
one 400 MW generator either at bus 18 or bus 21 with
other relatively large generating units causes load curtailment at the buses. Bus 18 has the lowest value of
adequacy because of many outage combinations of the connected generator with any other relatively large generator in the system. Bus 13 and Bus 20 have low values of
adequacy because of the outage of a generator at bus 23
with another larger generator outage in the system. The
results can be summarized as follows:

Line overload is not a major problem except for the
outages of a few lines in which case the connection between bus 6 and bus 10 is overloaded.

The contribution of line outages to reliability indices is negligibly small.

The major contribution to bus indices is due to generator deficiency in the system when large generating units are out.

Comparison Between The Methods

Both of the techniques described in this chapter are attempts to evaluate composite or bulk system adequacy. There are, however, conceptual differences in modeling and problem perception between the two techniques. In summary, some of the major differences in the models used, the risk indices obtained, the contingencies examined and the load alleviation techniques employed are shown in Table 3.18.

The results shown in Tables 3.15, 3.16 and Tables 3.17, 3.18 cannot be compared directly primarily because of the difference in load models used in the two studies. This difference is quite obvious. In order to recognize other differences and to facilitate a direct comparison, several specific conditions were examined as follows:

Case 1. The load was held constant at 2850 MW for the entire year and no planned maintenance was considered. The expected energy curtailed is 125,215 MWh using the ENEL approach and 145,819 MWh using the U of S approach.

The expected energy curtailed is higher in the U of S calculation due to load curtailment resulting from line overload alleviation arising from a.c. load flow representation. This is illustrated specifically in Cases 2 and 3.

Case 2. The load was held constant at 2850 MW for the entire year and all network components were in service. One 400 MW unit at bus 18 and one 197 MW unit at bus 13 were on forced outage. In the ENEL case, the overall available capacity is lower than the load to be

Table 3.18 - A comparison of the major factors

Factor	ENEL	U of S
Method	Simulation	Analytical
Risk index	Yearly curtailed energy subdivided according to cause	Range of load point and system indices
Situations examined	For a fixed year, 8760 hourly samples resulting from the random combination of any load level and any availability of system components	Predetermined contingency level up to four generating units and two transmission lines, at a specified load level
Load model	For the year considered any load level (8760 load levels)	A specified load level
Generation analysis	Planned maintenance of all units, predetermined for the year	Not included
Risk due to lack of generation	First component evaluated by curtailing load in accordance with a priority list	Curtailment by a firm/curtailable load policy and regional bounds
Initial dispatch	Priority list - based on running costs	Pre-specified
Network analysis	d.c. load flow, voltage	a.c. load flow, voltage
Load flow	limits not included	VAr limits included
Overload relief policy	d.c. load flow, overloads relieved using a "coefficient of influence" policy	Alleviation at close proximity points

supplied. Load shedding is necessary due to lack of generating capacity but there are no line overloads. In the U of S approach, the load is curtailed due to inadequate generation capacity in accordance with the prespecified policy. Again no line overloads occur.

Case 3. The load was held constant at 2850 MW for the entire year. One 400 MW unit at bus 18 and one 197 MW unit at bus 13 were on forced outage. The line between bus 2 and 6 was removed from service. The ENEL approach gives the same results as in Case 2. No overload conditions are determined by the d.c. load flow. In the U of S approach, load curtailment occurs at bus 6 due to overloading of the line between bus 6 and bus 10 in addition to load shedding due to generation inadequacy.

The three cases considered illustrate some of the fundamental differences between the two methods. In Case 1, the expected energy curtailed is basically the same and attributable largely to generation deficiencies. It should be appreciated that the load bus components of the curtailed energy will be very dependent on the load curtailment philosophy employed. The additional component in the U of S result arises as shown in Cases 2 and 3 due to load alleviation to remove difficulties perceived by the a.c. load flow.

CONCLUSIONS

This chapter has presented a comprehensive discussion of the effect of pertinent factors in the adequacy assessment of a composite power system. It has been noted that fast AC load flow techniques may be essential in certain adequacy studies. A description of a load curtailment philosophy that has been incorporated into a contingency enumeration approach is also given. The load curtailment philosophy described in this chapter is quite flexible. The load interruption can be localized in the neighborhood of a disturbance or it can be distributed throughout the system by assigning a proper load curtailment pass.

The inclusion of high level contingencies is necessary when calculating adequacy indices for relatively large power networks. This, however, results in a large computation time because the number of contingencies at higher levels becomes very large. In order to reduce the computation time, it is appropriate to incorporate only those credible outage events whose contribution to the adequacy indices cannot be ignored. A reduction in the computation time for the solution of generator outage contingencies is obtained by sorting identical units. This approach can result in a considerable saving in computation time, particularly when higher outage levels are considered. In order to include the contribution of outage events beyond the 4th level in the case of generating units and the 2nd level in the case of lines, the indices at the last level are modified such that the contribution of higher level outages is included without actually solving them.

The effect of higher level outage contingencies is not uniform at all the system buses. In the case of the RTS, calculation of the 4th level outage contingencies is necessary for both bus and system indices. The variation in the adequacy indices from one contingency level to another is also non-uniform for each bus. This variation depends upon the load curtailment philosophy and the relative location of the bus in the system.

The inclusion of dependent outages particularly those due to station originated outage events creates a significant increase in the load point and system adequacy indices, particularly at the lower system load levels. It is, therefore, necessary to examine these events prior to considering the inclusion of higher level independent outage events, particularly line outages. The total computation time does not increase significantly with the

addition of dependent outages but can increase tremendously with the inclusion of higher level independent
outages. The load at each bus, and hence the system
load, does not remain at a constant value throughout the
year. Generally, system load remains at its peak value
for only a short time. At lower load levels, the effect
of station originated outage events on the system adequacy may be comparable to that of independent outage
events. It is, therefore, very important to include the
effect of station originated events when calculating the
adequacy indices.

   This chapter has illustrated two conceptually different techniques for composite generation and transmission
system adequacy evaluation. The first approach using
Monte Carlo simulation is typical of the methods employed
by some utilities in Europe. The second approach uses a
contingency enumeration approach and is representative of
a number of techniques developed in North America and
elsewhere.

   The calculated results shown in this paper for the
RTS using the two methods are not directly comparable but
the differences are assignable. The major difference is
due to the load models employed. The solution techniques,
network representation and load curtailment policies
employed all introduce further assignable differences in
the calculated indices. In the case of the RTS, the load
reduction philosophy utilized in the case of line overload and inadequate installed capacity is a major factor.
The calculated individual load point adequacy indices
will vary widely with the load curtailment policy employed. The calculated indices are relative values which
respond to various factors that influence the adequacy of
the system. They can be used to consider planning alternatives and assess the sensitivity of the system to configuration changes.

REFERENCES

1. IEEE Committee Report, "Bibliography On The Application Of Probability Methods In Power System Reliability Evaluation," IEEE Trans. on Power Apparatus and Systems, PAS-91, 1972, pp. 649-660.
2. IEEE Committee Report, "Bibliography On The Application Of Probability Methods In Power System Reliability Evaluation, 1971-1977," IEEE Trans. on Power Apparatus and Systems, PAS-97, 1978, pp. 2235-2242.
3. Allan, R.N., Billinton, R. and S.H. Lee, "Bibliography On The Application Of Probability Methods In Power System Reliability Evaluation, 1977-1982," IEEE Trans. on Power Apparatus and Systems, PAS-103, 1984, pp. 275-282.
4. Mallard, S.A. and Thomas, V.C., "A Method For Calculating Transmission System Reliability," IEEE Transactions PAS-87, March 1968, pp. 824-833.
4. Billinton, R., "Composite System Reliability Evaluation," IEEE Transactions PAS-88, April 1969, pp. 276-80.
6. Billinton, R. and Bhavaraju, M.P., "Transmission Planning Using A Reliability Criterion - Part I - A Reliability Criterion," IEEE Transactions PAS-89, No. 1, January 1970, pp. 28-34.
7. Auge, J., Bergougnoux, J., Dodu, J.C. and Pouget, J., "Probabilistic Study Of A Transmission System Interconnection (Peru Model)," CIGRE paper 32-16, presented at the CIGRE Conference Paris, 1970 Session - 24 Aug. - 2 Sept.
8. Noferi, P.L. and Paris, L., "Quantitative Evaluation Of Power System Reliability In Planning Studies," IEEE Transactions Vol. PAS-91, No. 2, Mar./Apr. 1972, pp. 611-618.
9. Paris, L., Reggiani, F. and Valtorta, M., "The Study Of UHV System Reliability In Connection With Its Structure And Component Characteristics," CIGRE paper 31-14, presented at the CIGRE Conference Paris, 1972 Session - 28 Aug. - 6 Sept.
10. Bhavaraju, M.P. and Billinton, R., "Transmission System Reliability Methods," IEEE Transactions Vol. PAS-91, No. 2, Mar./Apr. 1972, pp. 628-637.
11. Bertoldi, O., Noferi, P.L. and Reggiani, F., "Improvement In Quantitative Evaluation Of Power System Reliability," IEEE Winter Power Meeting, New York, Jan./Feb. 1974, Paper No. C 74 137-6.
12. Noferi, P.L., Paris, L. and Salvaderi, L., "Monte Carlo Method For Power System Reliability Evaluation In Transmission And Generation Planning," Proceedings 1975 Annual Reliability and Maintainability Symposium, Washington, D.C., January 1975, Paper 1294, pp. 449-459, (IEEE Catalog No. 75-CHO-918-3 RQC).
13. Noferi, P.L. and Paris, L., "Effect Of Voltage And Reactive Power Constraints On Power System Reliability," IEEE Transactions, Vol. PAS-94, Mar./Apr. 1975,

pp. 482-490.

14. Dodu, J.C. and Merlin, A., "An Application Of Linear Programming To The Planning Of Large Scale Power Transmission Systems: The Mexico Model," Proceedings 5th PSCC, paper 22/9, Cambridge, England, September 1975.

15. Billinton, R., "Elements Of Composite System Reliability Evaluation," CEA Transactions, Vol. 15, Pt. 2, 1976, Paper No. 76-SP-148.

16. Dandeno, P.L., Jorgensen, G.E., Puntel, W.R. and Ringlee, R.J., "Program For Composite Bulk Power Electric System Adequacy Assessment," Transactions of the IEE Conference on Reliability of Power Supply Systems, IEE Conference Publication No. 148, February 1977.

17. IEEE Committee Report, "Reliability Indices For Use In Bulk Power Supply Adequacy Evaluation," IEEE Transactions, PAS-99, 1978, pp. 1097-1103.

18. Marks, G.E., "A Method Of Combining High Speed Contingency Load Flow Analysis With Stochastic Probability Methods To Calculate A Quantitative Measure Of Overall Power System Reliability," IEEE Paper A78 053-1.

19. Merlin, A and Oger, P.H., "Application Of A Variance Reducing Technique To A Monte Carlo Simulation Model Of Power Transmission Systems," pp. 3-35, 3-44, Workshop Proceedings: Power System Reliability - Research Needs and Priorities, EPRI Report WS77-60, October 1978.

20. Clements, K.A., Ejebe, G.C. and Wollenberg, B.F., "Linear Programming Network Flow Methods Applied To Bulk Power Supply Adequacy Assessment," IEEE Paper A78 062-2.

21. Mikolinnas, T.A. and Clements K.A., "Load Models For Bulk Power Supply Adequacy Assessment," IEEE Paper A78 232-1.

22. Billinton, R. and Medicherla, T.K.P., "Composite Generation And Transmission System Reliability Evaluation," IEEE Paper A78 237-0.

23. Garver, L.L., Van Horne, P.R. and Wirgau, K.A., "Load Supplying Capability Of Generation - Transmission Systems," IEEE Transactions, PAS-98, 1979, pp. 957-962.

24. Billinton, R., Medicherla, T.K.P. and Sachdev, M.S., "Adequacy Indices For Composite Generation And Transmission System Reliability Evaluation," IEEE Paper A79 024-1.

25. Atiyyah, I. and El-Abiad, A., "An Approach For Bulk Power System Reliability Evaluation," IEEE Paper A79 108-2.

26. Porretta, B., "Concepts For Composite System Reliability Applied To The Two-Area Planning Problem," IEEE Paper A79 109-0.

27. Billinton, R. and Medicherla, T.K.P., "Overall Approach To The Reliability Evaluation Of Composite

Generation And Transmission Systems," Proc. IEE, Part
C, 127, 1980, pp. 72-81.

28. Winter, W.H., "Measuring And Reporting Overall Reliability Of Bulk Electricity Systems," CIGRE, 1980,
paper 32-15.

29. Manzoni, G., Paris, L., Salvaderi, L. and Valtorta,
M., "Application Of Methods And Computing Programs To
System Planning. Part 1: Generation System, Part 2:
Transmission System," Journal of Electrical Power En-
ergy Systems, July 1980, Vol. 2, No. 3, pp. 147-158.

30. Billinton, R., Medicherla, T.K.P. and Sachdev, M.S.,
"Application Of Common-Cause Outage Models In Composite System Reliability Evaluation," IEEE Transactions, PAS-100, 1981, pp. 3648-3657.

31. Ranade, S.J. and Sullivan, R.L., "A Reliability
Analysis Technique For Bulk System Planning," IEEE
Transactions, PAS-100, 1981, pp. 3658-3665.

32. Billinton, R. and Medicherla, T.K.P., "Station
Originated Multiple Outages In The Reliability Analysis Of A Composite Generation And Transmission System," IEEE Transactions, PAS-100, 1981, pp. 3870-
3878.

33. Van Horne, P.R. and Schoenberger, C.N., "TRAP: An
Innovative Approach To Analysing The Reliability Of
Transmission Plans," IEEE Transactions, PAS-101,
1982, pp. 11-16.

34. Mikolinnas, T.A., Puntel, W.R. and Ringlee, R.J.,
"Application Of Adequacy Assessment Techniques For
Bulk Power Systems," IEEE Transactions, PAS-101,
1982, pp. 1219-1228.

35. Endrenyi, J., Albrecht, P.F., Billinton, R., Marks,
G.E., Reppen, N.D. and Salvaderi, L., "Bulk Power
System Reliability Assessment - Why And How?   Part
I: Why," IEEE Transactions, PAS-101, 1982, pp.
3439-3445.

36. Endrenyi, J., Albrecht, P.F., Billinton, R., Marks,
G.E., Reppen, N.D. and Salvaderi, L., "Bulk Power
System Reliability Assessment - Why And How?   Part
II: How?," IEEE Transactions, PAS-101, 1982, pp.
3446-3456.

37. "Bulk Power Area Reliability Evaluation Considering
Probabilistic Transfer Capability," PJM Transmission
Reliability Task Force, IEEE Transactions, PAS-101,
1982, pp. 3551-3562.

38. Allan, R.N. and Adraktas, A.N., "Terminal Effects And
Protection System Failures In Composite System Reliability Evaluation," IEEE Transactions, PAS-101,
1982, pp. 4557-4562.

39. Billinton, R. and Tatla, J., "Composite Generation
And Transmission System Adequacy Evaluation Including
Protection System Failure Modes," IEEE Transactions
PAS-102, No. 6, June 1983, pp. 1923-1930.

40. Meliopoulos, A.P., Bakirtzis, A.G. and Kovacs, R.,
"Power System Reliability Evaluation Using Stochastic
Load Flows," IEEE Transactions PAS-103, No. 5, May

1984, 1084-1091.

41. Clements, K.A., Lam, B.P., Lawrence, D.J. and Reppen, N.D., "Computation Of Upper And Lower Bounds On Reliability Indices For Bulk Power Systems," PAS-103, No. 8, August 1984, pp. 2318-2325.

42. Billinton, R. and Kumar, S., "Adequacy Evaluation Of A Composite Generation And Transmission System - A Comparative Study Of Existing Programs," CEA Trans., Vol. 24, 1985.

43. Billinton, R., Vohra, P.K. and Kumar, S., "Effect Of Station Originated Outages In A Composite System Adequacy Evaluation Of The IEEE Reliability Test System," IEEE Transactions PAS-104, No. 10, October 1985, pp. 2649 - 2656.

44. Cunha, S.H.F., Pereira, M.V.F., Pinto, L.M.V.G. and Oliveira, G.C., "Composite Generation And Transmission Reliability Evaluation In Large Hydroelectric Systems," IEEE Transactions PAS-104, No. 10, October 1985.

45. Lauby, M.G., Doudna, J.H., Polesky, R.W., Lehman, P.J. and Klempel, D.D., "The Procedure Used In The Probabilistic Transfer Capability Analysis Of The MAPP Region Bulk Transmission System," IEEE Trans. PAS-104, No. 11, November 1985, pp. 3013-3019.

46. Salvaderi, L. and Billinton, R., "A Comparison Between Two Fundamentally Different Approaches To Composite System Reliability Evaluation," IEEE Transactions PAS-104, No. 12, December 1985, pp. 3486-3492.

47. Billinton, R. and Kumar, S., "A Comparative Study Of System Versus Load Point Indices For Bulk Power Systems," IEEE 1986 Winter Power Meeting, Paper No. 86 WM 045-9.

48. Billinton, R. and Kumar, S., "Pertinent Factors In The Adequacy Assessment Of A Composite Generation And Transmission System," CEA Trans., Vol. 25, 1986.

49. Meliopolis et al., "Bulk Power Reliability Assessment With The RECS Program," Proceedings PICA, May 1985, pp. 38-46.

50. EPRI Report, "RP 1530-Transmission System Reliability Methods," Project RP 1530-1, EL2526, July 1982.

51. EPRI Report, "RP 1353-Reliability Indexes For Power Systems," Project RP 1353-1 EL1773, March 1981.

52. Billinton, R., "Power System Reliability Evaluation," Gordon and Breach Science Publishers, New York, 1970.

53. Billinton, R., Ringlee, R.J. and Wood, A.J., "Power System Reliability Calculations," MIT Press, Mass., 1973.

54. Billinton, R. and Allan, R.N., "Reliability Evaluation Of Power Systems," Longman, London (England)/ Plenum Publishers, New York, 1984.

55. Landgren, G.L. and Anderson, S.W., "Data Base For EHV Transmission Reliability Evaluation," IEEE Trans. PAS-100, 1981, pp. 2046-2058.

56. EPRI Workshop, "Power System Reliability - Research Needs And Priorities," EPRI Report, WS77-60, October

1978.
57. Allan, R.N. and Ochoa, J.R., "Modeling And Assessment
    Of Station Originated Outages For Composite System
    Reliability Evaluation," IEEE Winter Power Meeting,
    New Orleans, 1987, paper 87 WPM 016-9.
59. IEEE Committee Report, "IEEE Reliability Test Sys-
    tem," IEEE Transactions PAS-98, 1979, pp. 2047-2054.
60. Allan, R.N., Billinton, R., Shahidehpour, S.M. and
    Singh, C., "Bibliography On The Application Of Prob-
    ability Methods In Power System Reliability Evalua-
    tion, 1982-87," IEEE Winter Power Meeting, New York,
    February 1988.

CHAPTER 4

DISTRIBUTION SYSTEM ADEQUACY EVALUATION

INTRODUCTION

Reliability assessment of a distribution system is usually concerned with the system performance at the customer end, i.e. at the load points. The basic indices [1] normally used to predict the reliability of a distribution system are: load point failure rate, average outage duration and annual unavailability. The basic indices are important from an individual customer's point of view but they do not provide an overall appreciation of the system performance. An additional set of indices can be calculated using these three basic indices and the number of customers/load connected at each load point in the system. Most of these additional indices are weighted averages of the basic load point indices. The most common additional or system indices [1] are; System Average Interrruption Frequency Index (SAIFI), System Average Interruption Duration Index (SAIDI), Customer Average Interruption Duration Index (CAIDI), Average Service Availability Index (ASAI), Average Service Unavailability Index (ASUI), Energy Not Supplied (ENS) and Average Energy Not Supplied (AENS). These system indices are also calculated by a large number of utilities from system interruption data and provide valuable indications of historic system performance. The ability to calculate the same basic indices for future performance as is used to measure past performance is an important consideration. This chapter illustrates the calculation of the system

performance indices from historical customer interruption
data and the determination of the system predictive indices using component outage data.

Reliability indices of a distribution system are
functions of component failures, repairs and restoration
times which are random by nature. The calculated indices
are therefore random variables and can be described by
probability distributions. Conventional distribution
system reliability evaluation is normally concerned only
with the average or mean values of the calculated indices. There is however considerable additional information, which could assist the system planner, designer and
manager in the distributions associated with each calculated index.

This chapter illustrates two approaches which have
been used to examine the probability distributions associated with distribution system reliability indices. The
first technique utilizes Monte Carlo simulation, while
the second is a direct analytical approach utilizing the
raw moments of the data and the Pearson distributions.
Both techniques are illustrated by application to a distribution configuration.

HISTORICAL SYSTEM PERFORMANCE ASSESSMENT

System Data

The calculation of the basic system performance indices can be easily illustrated by a small example. Consider the system data shown in Table 4.1. This is a relatively small system consisting of six main feeders with
55,000 customers. The assumed customer interruption data
[1] for the year are given in Table 4.2 and apply to sustained interruptions. A similar calculation could be
performed for momentary interruptions.

Table 4.1 – System data

Bus	No. of customers served by feeders from bus
A	5,000
B	15,000
C	10,000
D	10,000
E	7,000
F	8,000
Total	55,000

Table 4.2 – Customer interruption data

Interruption case	Customer interruptions		Duration (hours)
1	A	5,000	1.0
	D	1,000*	0.2
2	C	5,000	2.0
3	B	4,000	0.5
4	D	2,000*	1.75

Total interruptions	17,000
Customers affected	16,000*

*Interruption cases 1 and 4 involve the same customers served by feeders from D bus. Hence Interruption Case 4 adds 1,000 new affected customers to the list of affected customers. The condition that each customer affected is counted only once regardless of the number of interruptions experienced requires that a memory be included in the data collection process. This is somewhat difficult to accomplish and is not normally done by most utilities.

System Indices

The indices of SAIFI, CAIFI, SAIDI, CAIDI and ASAI
are defined and calculated as follows.

System Average Interruption Frequency Index – SAIFI.
This index is defined as the average number of interruptions per customer served per time unit. It is estimated
by dividing the accumulated number of customer-interruptions in a year by the number of customers served. The
evaluation of SAIFI is shown in Table 4.3.

Table 4.3 – Evaluation of SAIFI

Interruption case	Customers interrupted	
1	A	5,000
	D	1,000
2	C	5,000
3	B	4,000
4	D	2,000

Total customer interruptions = 17,000

System average interruption frequency index =

$$\text{SAIFI} = \frac{17,000}{55,000} = \frac{0.31}{\text{customer} - \text{yr}}$$

**Customer Average Interruption Frequency Index –
CAIFI.** This index is defined as the average number of
interruptions experienced per customer affected per time
unit. It is estimated by dividing the number of customer
interruptions observed in a year by the number of customers affected. Each customer affected is counted only
once regardless of the number of interruptions that
customer may have experienced during the year.
Evaluation of CAIFI is shown in Table 4.4.

Table 4.4 – Evaluation of CAIFI

Interruption case		Customers Interrupted
1	A	5,000
	D	1,000*
2	C	5,000
3	B	4,000
4	D	2,000*

Total customer interruptions = 17,000

Total customers* affected   = 16,000

Customer average interruption frequency index =

$$\text{CAIFI} = \frac{17,000}{16,000} = \frac{1.06 \text{ interruptions/yr}}{\text{customers affected/yr}}$$

System Average Interruption Duration Index – SAIDI. This index is defined as the average interruption duration for customers served during a year. It is determined by dividing the sum of all customer sustained interruption durations during the year by the number of customers served during the year. Evaluation of SAIDI is shown in Table 4.5.

Table 4.5 – Evaluation of SAIDI

Interruption case	Customer interrupted		Duration minutes	Customer-minutes
1	A	5,000	60	300,000
	D	1,000	12	12,000
2	C	5,000	120	600,000
3	B	4,000	30	120,000
4	D	2,000	105	210,000

Customer interruptions = 17,000

Cumulative customer–minutes interruption = 1,242,000

System average interruption duration index =

$$\text{SAIDI} = \frac{1,242,000}{55,000} = \frac{22.58 \text{ minutes}}{\text{system customer}}$$

Customer Average Interruption Duration Index – CAIDI.
This index is defined as the interruption duration for
customers interrupted during a year. It is determined by
dividing the sum of all customer sustained interruption
durations during the specified period by the number of
sustained customer interruptions during the year.

Customer average interruption duration index =

$$\text{CAIDI} = \frac{1,242,000}{17,000} = \frac{73 \text{ minutes}}{\text{customer interrupted}}$$

Average Service Availability Index – ASAI. This is
the ratio of the total number of customer hours that service was available during a year to the total customer
hours demanded.   Customer hours demanded are determined
as the twelve-month average number of customers served
times 8760.   The complementary value to this index, i.e.
the Average Service Unavailability Index (ASUI) may also
be sometimes used.

Average service availability index =

$$\text{ASAI} = \frac{(55,000 \times 8760) - (1,242,000/60) - 0.999957}{55,000 \times 8760}$$

$$\text{ASUI} = 1 - \text{ASAI} = 0.000043$$

Statistics on Canadian utility performance using
SAIFI, SAIDI, CAIDI and ASAI have been collected and published by the Distribution System Reliability Committee
of the Canadian Electrical Association for many years.
Individual utility statistics are presented in Reference
[2].   Table 4.6 shows global Canadian statistics for the
1984 and 1985 annual periods.

Table 4.7 shows the variation in ASAI for the 1962 –
85 period. This statistic has been very popular in Canada
and was in fact the basic statistic used in the early
data collection systems.

Table 4.6 - Canadian utility performance data

	1984	1985
SAIFI	2.73	2.48
SAIDI	4.93	4.11
CAIDI	1.81	1.66

Table 4.7 - Canadian utility performance data

Canadian utility performance data

Year	ASAI	Year	ASAI	Year	ASAI
1962	0.999780	1970	0.999592	1978	0.999598
1963	0.999700	1971	0.999632	1979	0.999617
1964	0.999698	1972	0.999705	1980	0.999561
1965	0.999707	1973	0.999018	1981	0.999557
1966	0.999737	1974	0.999799	1982	0.999557
1967	0.999701	1975	0.999602	1983	0.999565
1968	0.999645	1976	0.999516	1984	0.999437
1969	0.999514	1977	0.999510	1985	0.999531

It can be seen that on the average, Canadian consumers have received electric energy service better than 99.9% of the time over this period. Not all Canadian utilities collect service continuity statistics. The data in Tables 4.6 and 4.7 cover approximately 60% of the metered load points in Canada.

The assessment of past performance is extremely important. This must be done on a quantitative basis and can provide a useful range of indices. It can also be argued that it is somewhat difficult, if not impossible, to consider future performance of a system without having a consistent awareness of how the system has performed in the past.

BASIC DISTRIBUTION SYSTEMS

A distribution circuit normally uses primary or main feeders and lateral distributors. A main feeder originates at the substation and passes through the major load centers. The individual load points are connected to the

main feeder by lateral distributors with distribution
transformers at their ends. A main feeder is constructed
using single, parallel or meshed circuits. Many distribution systems used in practice have a single circuit
main feeder and are defined as radial distribution systems. There are also many systems which, although constructed using meshed circuits, are operated as radial
systems using normally open switches in the meshed circuit. The main feeder, in some cases, may have branches
to reach the widely distributed areas. Radial systems
are popular due to their simple design and generally low
cost. These systems have a set of series components between the substation and the load points. The failure of
any of these components causes outage of the load
point(s). The outage duration and the number of customers affected due to a component failure are reduced by
using extensive protection and sectionalizing schemes.
The sectionalizing equipment provides a convenient means
of isolating the faulted section. The supply can then be
restored to the healthy sections, maintaining the service
to some of the load points, while the faulted component
is being repaired. The time taken by this type of isolation and switching action is referred to in this chapter
as restoration time. In some systems, there is a provision for alternative supply in the case of a failure.
This alternative source is used to supply that section of
the main feeder which becomes disconnected from the main
supply after the faulted section has been isolated. The
alternative supply, however, may not always be available
for some reason. The probability associated with the
availability of the alternative supply must therefore, be
included in the analysis [1]. Fuse equipment is usually
provided on the lateral distributor. Faults on the lateral distributor or in the distribution transformer are
normally cleared by this equipment and therefore, service
on the main feeder is maintained. If the fuse fails to

clear the fault for some reason, the circuit breaker or
the backup fuse on the main feeder acts to clear the
fault. The faulted lateral distributor is then isolated
and the supply is restored to the rest of the system by
closing the circuit breaker. The reliability analysis
must therefore, include the probability associated with
the successful operation of the fuse [1].

DISTRIBUTION SYSTEM RELIABILITY ASSESSMENT

Base Case Analysis

The basic indices normally used to predict the reliability of a distribution system are: Average Load Point
Failure Rate, Average Load Point Outage Duration, and Average Annual Load Point Outage Time. The system performance indices, SAIDI, SAIFI, CAIDI and others can also be
calculated directly from the three basic predictive indices. A relatively simple calculation of reliability and
performance indices is presented for the example radial
system shown in Figure 4.1.

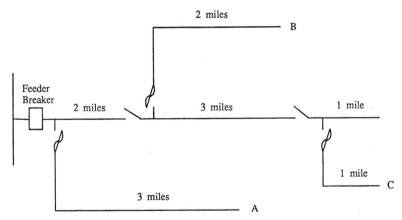

Figure 4.1 - Simple radial configuration

In this simple configuration, all switches are normally closed and the customer load points A, B, C are

supplied from the primary main by fused laterals. The
feeder breaker and the substation supply bus are assumed
to be fully reliable. The individual component data are
as follows:

Primary main   0.10 f/circuit mile/yr
                        3.0 hr average repair time

Primary lateral   0.25 f/circuit/mile/yr
                        1.0 hr average repair time

Manual sectionalizing time for any switching action = 0.5
hr.

The simplest approach is to perform a failure modes
and effect analysis in a table form utilizing the following basic equations [1]:

$$\lambda_s = \Sigma \lambda_i \qquad \text{f/yr}$$

$$r_s = \frac{\Sigma \lambda_i r_i}{\Sigma \lambda_i} \qquad \text{hr/failure}$$

$$U_s = \lambda_s \cdot r_s \quad \text{hr/yr}$$

This procedure is shown in Table 4.8 and the results are
summarized in Table 4.9.

It can be seen that load point C, despite being at
the extremity of the primary main, has the lowest failure
rate due to its relatively short primary lateral. It has
the longest average restoration time, however, due to the
fact that all restoration is by repair rather than by
isolation. In the case of load point A, any failures on
the primary main other than on the initial 2 mile section
involve restoration by switching rather than repair.
There are many configurations particularly in rural locations which have a topology similar to that shown in Figure 4.1. The results shown in Table 4.9 can be used to
obtain the standard performance indices. Assume that
there are 250, 100 and 50 customers respectively at load
points A, B and C.

Table 4.8 - Case 1 - basic calculations

	Load Point A			Load Point B			Load Point C		
Component	λ f/yr	r hr	U hr/yr	λ f/yr	r hr	U hr/yr	λ f/yr	r hr	U hr/yr
Primary main									
2 m section	0.2	3.0	0.6	0.2	3.0	0.6	0.2	3.0	0.6
3 m section	0.3	0.5	0.15	0.3	3.0	0.9	0.3	3.0	0.9
1 m section	0.1	0.5	0.05	0.1	0.5	0.05	0.1	3.0	0.3
Primary lateral									
3 m section	0.75	1.0	0.75	-	-	-	-	-	-
2 m section	-	-	-	0.5	1.0	0.5	-	-	-
1 m section	-	-	-	-	-	-	0.25	1.0	0.25
	1.35	1.15	1.55	1.1	1.86	2.05	0.8	2.41	2.05

Table 4.9 - Summary of Case 1

		A	B	C
λ	- f/yr	1.35	1.10	0.85
r	- hr/failure	1.15	1.86	2.41
U	- hr/year	1.55	2.05	2.05

Total number of customers served = 400

Annual Customer Interruptions =

$(250)(1.35) + (100)(1.1) + (50)(0.85) = 490$

Customer Interruption Duration =

$(250)(1.55) + (100)(2.05) + (50)(2.05) = 695$

System Average Interruption Frequency Index = SAIFI

$$= \frac{\text{total number of customer interruptions}}{\text{total number of customers served}}$$

$$\text{SAIFI} = \frac{490}{400} = 1.23$$

System Average Interruption Duration Index = SAIDI

$$= \frac{\text{sum of customer interruption durations}}{\text{total number of customers}}$$

$$\text{SAIDI} = \frac{695}{400} = 1.74$$

Customer Average Interruption Duration Index = CAIDI

$$= \frac{\text{sum of customer interruption duration}}{\text{total number of customers interrupted}}$$

$$\text{CAIDI} = \frac{695}{490} = 1.42$$

Average Service Availability Index = ASAI

$$= \frac{\text{customer hours of available service}}{\text{customer hours demanded}}$$

$$\text{ASAI} = \frac{(400)(8760) - 695}{(400)(8760)} = 0.999802$$

These calculated values can be compared with measured
values or, if available, with standard indices for the
system to determine if the configuration shown in Figure
4.1 meets the system requirement.

Sensitivity Analyses

It may also be possible to restore service to this
system by back feeding from another adjacent circuit.
This configuration is shown in Figure 4.2.

Table 4.10 shows the effect of this alternative supply on the calculated reliability indices using an average switching time of 1 hour for the alternative supply.
The results are summarized in Table 4.11.

It can be seen that the load point failure rates are
not affected by the ability to backfeed from an alternative supply. This will apply in all cases in which the

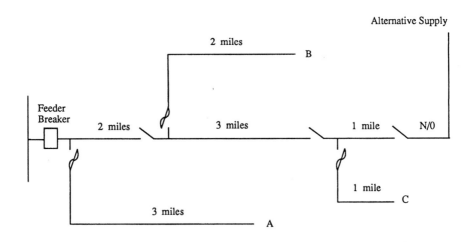

Figure 4.2 — Manually sectionalized primary main with alternative supply

Table 4.10 — Case 2 — effect of alternative supply

	Load Point A			Load Point B			Load Point C		
Component	λ f/yr	r hr	U hr/yr	λ f/yr	r hr	U hr/yr	λ f/yr	r hr	U hr/yr
Primary main									
2 m section	0.2	3.0	0.6	0.2	1.0	0.2	0.2	1.0	0.2
3 m section	0.3	0.5	0.15	0.3	3.0	0.9	0.3	1.0	0.3
1 m section	0.1	0.5	0.05	0.1	0.5	0.05	0.1	3.0	0.3
Primary lateral									
3 m section	0.75	1.0	0.75	---			---		
2 m section				0.5	1.0	0.5	---		
1 m section				---			0.25	1.0	0.25
	1.55	1.15	1.55	1.1	1.5	1.65	0.85	1.24	1.05

restoration of service is done manually. If automatic
switching is used and the customer outage time is considered to be so short that the event is not classed as a
failure, then the overall failure rate will be reduced to
a value closely related to the primary lateral value.
This assumes that the automatic sectionalizing and ser-

Table 4.11 – Summary of Case 2

	A	B	C
λ – f/yr	1.35	1.1	0.85
r – hr/failure	1.15	1.5	1.24
U – hr/yr	1.55	1.65	1.05

Overall configuration indices

SAIFI   =   1.23

SAIDI   =   1.51

CAIDI   =   1.23

ASAI   =   0.999827

vice restoration has a high probability of successful operation. The ability to backfeed has a pronounced effect
on the length of the interruption particularly for those
customers at the extremities of the primary main. This
effect could be reduced considerably if the ability to
backfeed is conditionally dependent upon the loading condition in the alternative supply. The restoration times
in Table 4.10 for load points B and C can be modified to
reflect the probability of being able to supply these
loads from the alternative supply. This effect is shown
as Case 3 in Table 4.12 using a transfer probability of
0.5.

It can be seen from the cases studied that the load
point failure rates are dependent upon the components exposed to failure and the degree of automatic isolation of
a failed component in the network. This effect can be
easily seen in the network of Figure 4.1. If each lateral is solidly connected to the primary main, all load
points will have the same failure rate, as any fault will
result in the feeder breaker tripping. These results are
shown as Case 4 in Table 4.12.

The results shown in Table 4.9 illustrate the effect
of perfect isolation arising from a failure on a primary

Table 4.12 - Results for various system conditions

Summary

	Case				
	1	2	3	4	5
Load point A					
λ f/yr	1.35	1.35	1.35	2.10	1.425
r hr	1.15	1.15	1.15	0.92	1.114
U hr/yr	1.55	1.55	1.55	1.93	1.5875
Load point B					
λ f/yr	1.10	1.10	1.10	2.10	1.20
r hr	1.86	1.50	1.68	1.39	1.75
U hr/yr	2.05	1.65	1.85	2.93	2.10
Load point C					
λ f/yr	0.85	0.85	0.85	2.10	0.975
r hr	2.41	1.24	1.82	1.57	2.17
U hr/yr	2.05	1.05	1.55	3.30	2.1125
System indices					
SAIFI	1.23	1.23	1.23	2.10	1.31
SAIDI	1.74	1.51	1.63	2.35	1.78
CAIDI	1.42	1.23	1.33	1.12	1.36
ASAI	0.999802	0.999827	0.999814	0.999732	0.999797

Case	Condition
1.	Base case shown in Figure 4.1.
2.	System shown in Figure 4.2, alternative supply average switching time of 1 hr.
3.	As in Case 2, conditional load transfer probability of 0.5.
4.	As in Case 1, solidly connected laterals.
5.	As in Case 1, probability of successful lateral fault clearing = 0.9.

lateral. The probability associated with successful isolation of a primary lateral fault will depend upon the design of the protection coordination scheme and on the operation and maintenance of the scheme. Table 4.12 also shows the results for Case 5 for which the probability of successful isolation of a primary lateral fault is 0.9.

DISTRIBUTIONAL CONSIDERATIONS

Concepts Of Distributions

Conventional reliability analyses are normally only concerned with the expected or average value of the particular measure of reliability. Little consideration has been given in the past to the variation of that measure about its mean. For example, when the frequency of failures at a load point is predicted, only the average value of that quantity is typically calculated. The probability that the load point will suffer a specified number of failures in a year is not normally considered. Similarly, the expected values of the duration indices are determined but the probabilities of various durations are not calculated. The mean values are extremely useful and are the primary indices of load point adequacy. There is, however, an increased awareness of the need for information related to the variation of the reliability measures around their means.

Probability distributions provide a practical vehicle to describe the variation of reliability measures about their means. One of the approaches taken to determine these distributions is to perform probabilistic (Monte Carlo) simulations of typical radial distribution systems. Load point index distributions are dependent not only on combinations of component outages but also on system configurations and restoration activities. With the increasing emphasis that utilities are placing on data collection, it is possible that in the near future more statistical data on load point interruptions will be available. Not only can simulation studies provide useful information before comprehensive historical data is available but they can provide information that would not otherwise be possible to obtain, such as information concerning the effects of specific system configurations.

Studies examining the distributions associated with the basic reliability indices indicate that the load

point failure rate is approximately Poisson distributed
[3,4]. The failure rate probabilities can be readily obtained using the mean load point failure rate, since the
Poisson distribution is a single-parameter function. This
information can be calculated directly using the Poisson
equation or from a set of published graphs [1].

It has been noted that if the restoration times can
be assumed to be exponentially distributed, the load
point outage duration is approximately gamma distributed
and the desired probability information can be readily
calculated [3,4]. There are, however, many distribution
systems for which the gamma distribution does not adequately describe the load point outage duration. These
systems are ones in which some of the restoration times
may be better described by non-exponential distributions,
e.g. lognormal repair or manual sectionalizing times.
Studies carried out at the University of Saskatchewan indicate that when the restoration times are assumed to be
non-exponential then the interruption duration can not
generally be represented by a gamma distribution [4,5,6].
In this case, obtaining the desired probability information is more difficult. Distributions for the Annual
Load Point Outage Time, SAIDI, SAIFI, and CAIDI indices
can also not readily be represented by common distributions.

Probabilistic Simulation

Method. A program has been developed at the University of Saskatchewan to simulate the performance of any
N-section radial distribution system with loads connected
to laterals or directly to the primary mains. Any combination of exponential, normal, lognormal, and gamma distributions can be used to simulate the failure, repair,
manual sectionalizing, alternative supply and fuse times.
Costs of each interruption can also be calculated from 1
minute, 20 minute, 1 hour, 4 hour and 8 hour cost data

such as that described in Chapter 5. The program outputs
for each load point: the mean, standard deviation, and
distribution histogram of the annual interruption time,
interruption duration, annual interruption frequency, and
annual interruption cost. It provides similar outputs
for the entire system, in terms of SAIFI, SAIDI, CAIDI,
cost per interruption, and annual interruption cost.
Studies have been performed on the 6 section example system of Figure 4.1 and on larger systems.

Load Point Failure Rates. The studies indicate that
the load point failure rate is reasonably described by
the Poisson distribution with a Chi-squared level of significance = 0.1 [4,5,6]. This result is in agreement with
theoretical considerations and a previous study by Patton
[3].   As only one parameter is required to describe the
Poisson distribution, i.e. the expected annual failure
rate, the distributional information can be obtained with
minimal extra effort.

Figure 4.3 presents the distributions associated with
the failure rates of load points A, B and C for Cases 2
and 4 in Table 4.12. The distributions are noticeably
different for the three load points. At load point A,
years with one failure occur most frequently while at
load point C, years with no failures occur most frequently.   Concurrent with an increase in the average
failure rate, the shape of the distribution varies significantly and the individual failure rate probabilities
increase in a non-linear fashion.

Load Point Outage Durations. Patton [3] noted that
if the repair and other restoration times can be assumed
to be exponentially distributed, the load point outage
duration can be approximated by a gamma distribution.
This distribution can take on many different shapes, becoming more spread out, or more peaked, depending on the

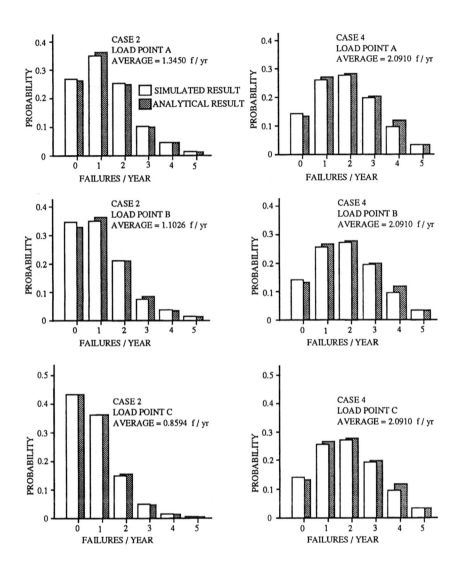

Figure 4.3 - Load point failure rate distributions

average outage durations and the average restoration durations. If it can be assumed that the restoration times
are exponentially distributed, the outage duration probabilities can be readily calculated from the gamma distribution.   In many practical systems, the restoration

times cannot be assumed to be exponentially distributed. As an example, it is often unrealistic to assume that the probability of a repair or restoration increases as the duration approaches zero. Restoration times may be better described by non-exponential distributions, e.g. lognormal repair times. The studies carried out indicate that when the restoration times are assumed to be non-exponential, the load point outage duration cannot generally be represented by a gamma distribution.

It should be emphasized that the average values of the load point outage duration indices are not affected by the underlying distributions. The averages, such as those calculated for the example system of Figure 4.1, can have any set of distributions associated with them.

System Indices. System performance indices and load point reliability indices are not normally compared or simultaneously calculated. Performance and reliability indices are in reality related by the fact that both are based on a common set of data. The distributions of both indices are dependent on the distributions of the component failure rates and the restoration activities. The SAIFI and SAIDI average indices are independent of the underlying distributions but the CAIDI average index is distributionally dependent.

Figure 4.4 depicts distributions resulting from simulations of the basic example system of Figure 4.1. The SAIDI distribution is dependent only on the distributions of the restoration times. The number of customers at each load point and the average failure rates are weighting factors that are independent of the associated distributions. The SAIDI distributions of Figure 4.4 are similar to the Annual Interruption Time distributions. In a large system, the resemblance tends to decrease because of the averaging effect of the large number of load points which are aggregated. In this small system, the number of years

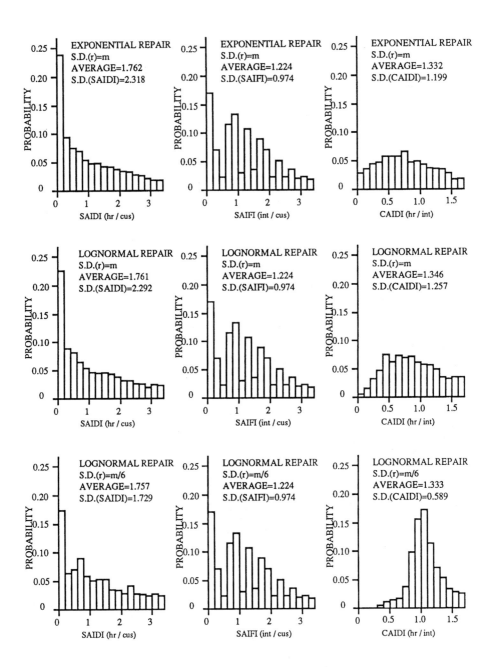

Figure 4.4 – Distribution of the system performance indices

with a SAIDI equal to zero is relatively high. This might
be expected in a small or moderate size system. In the
case of a large system or region it is very likely that
there would be at least one interruption.

The SAIFI distributions were found to be identical
for exponential and lognormal simulations because SAIFI
is only dependent on the component failure time distributions, which do not vary with the simulation runs, and on
the number of customers served at each load point. Similar to the discussion concerning the SAIDI distributions,
the probability of SAIFI equalling zero diminishes in
large systems. The result is that the distribution looks
less like an exponential one and more like one with a
mode about the average.

The CAIDI distributions are non-linearily related to
both the failure and restoration times. This and other
comparisons indicate that the standard deviations of the
underlying distributions can affect the shape of the final index distribution as much as or more than the actual
form chosen for the underlying distributions. For large
systems, the CAIDI distribution also tends to cluster
more around the mean.

Direct Analytical Evaluation

A computer program to perform predictive reliability
assessment studies on radial systems of the size and complexities of actual systems, has been developed. The program can analyze a wide range of radial distribution systems differing in configuration, operating procedure,
protection schemes and the modes of restoration. The
analysis can be done by knowing the system configuration
and having familiarity with its operation. Sensitivity
analysis with respect to the input parameters can also be
done using the program. This program has been utilized
to analyze a number of practical distribution configurations.

A general analytical approach has been developed to determine approximate information in the form of percentiles to describe the probability distributions of the reliability indices [7]. The details of this analytical approach are given in Appendix 4. The following is a brief descriptive summary.

A reliability index can be expressed as:

$$Z = f(X_1, X_2, X_3, \ldots)$$

where X_1, X_2, X_3, ... are the random variables, which denote the parameters related to the component performance and the system operation. The variable Z is a random variable because it is a function of random variables. The function f takes a form which depends on the system configuration and the reliability index represented by the function. The objective is to determine the probability distribution of the random variable Z if the function f and the probability distributions of the random variables X_1, X_2, X_3, ... are known. Direct analytical methods are available for obtaining the exact form of some simple algebraic functions of random variables. These methods do not provide solutions for all the types of probability distributions which are usually used to represent random variables. The reliability indices are intricate functions of random variables such as component repair times, restoration times etc., which can assume a wide range of probability distribution forms. It may, therefore, be difficult, unfeasible or perhaps even impossible to obtain, in exact form, the probability distributions of the reliability indices. An approximate solution is, however, possible. The detailed equations are presented in Appendix 4.

The analysis requires three major steps:

(a) Step 1 - the first four raw moments of component failure and repair times and the system restoration times are determined,

(b) Step 2 - the average value and the second, third and

fourth central moments of the reliability indices are
evaluated using the moments obtained in Step 1 and
the information regarding the system configuration,
(c) Step 3 - the Pearson Method is utilized to evaluate
the approximate percentiles of the reliability indices. The Pearson Method approximates the probability
distribution of a random variable by utilizing moments.

The solution obtained is approximate because a probability distribution is not fully described by the first
four moments. This approximation has, however, been found
to give good results. The Pearson technique can provide
an analytical expression for the approximate probability
distribution. A table [8] has been published to directly
obtain the percentiles of random variables. A computer
program has been developed utilizing the analytical approach to determine the percentiles of load point outage
duration, annual unavailability and the following system
indices: SAIFI, SAIDI, ASAI, ASUI, ENS, AENS. The number of load point interruptions in a year follows a Poisson distribution and therefore only the average value is
evaluated for this index. The repair times may be specified to have any of the standard distributions. The failure times are assumed to be exponentially distributed.

It is observed that in some cases the percentiles of
the reliability indices cannot be estimated because the
parameters fall out of the range of the tables used for
the Pearson approximation of the probability distributions of these indices. This happens in cases where the
probability distribution of an index is not unimodal.
The probability distributions associated with the component repair times and the system restoration times are
used as input parameters in the analysis. In some cases
only a set of observations on these variables may be
available instead of the exact analytical form of the
probability distributions followed by them. In these

cases it is not necessary to find a probability distribution which fits the observed data. The analytical techniques to obtain the moments of the reliability indices
utilize the first four raw moments of the input variables. The sample values of these moments can be directly obtained from the observed data and utilized in
the analysis.

PRACTICAL APPLICATIONS

Sample System

The RTS does not include distribution system components and therefore cannot be used to illustrate the concepts contained in this chapter. The Annaheim distribution system of the Saskatchewan Power Corporation has
been used as the basic configuration for the study. It is
a totally underground distribution system extending from
St. Gregor (south end) to Annaheim (north end). The configuration of the system is shown in Figure 4.5. The
system supplies electric power to twenty-four farms, using a branched main feeder. The primary feeder uses a
14.4 kV supply which can be connected to the system from
both the ends. The south end is used as the main supply
while the north end provides an alternative supply point
for emergency conditions. The failure rates and repair
times of the components, the system restoration times and
the probabilities associated with the fuse operation and
availability of the alternative supply are normally the
input parameters for the reliability analyses of a distribution system. These parameters are determined on the
basis of historical data. In cases where sufficient data
are not available to evaluate the parameters, sensitivity
analyses can be done to determine the effect on the reliability indices by individually varying each input parameter. The base parameters of the sample system are as
follows.

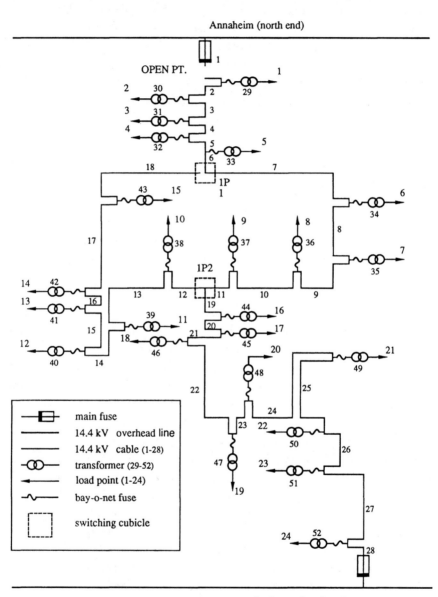

Figure 4.5 - Sample system

Input parameters:

    failure rate of cable sections = 0.017 f/yr km,

    repair time of cable sections = 8 hr,

    failure rate of transformers = 0.005 f/yr,

    replacement time of transformers = 48 hr,

    probability of fuses to clear the fault = 0.2,

    probability of alternative supply availability = 0.2.

Reliability Indices

    Table 4.13 shows the calculated indices for three load points and the overall system indices.

Table 4.13 - Basic reliability indices

Load point	Failure rate f/yr	Repair time hr	Unavailability hr/yr
3	0.7753	7.34	5.69
12	0.7753	6.34	4.41
18	0.7753	5.58	4.32

SAIFI = 0.7753 interruptions/system customer,

SAIDI = 4.561 hr/system customer,

CAIDI = 5.883 hr/customer interrupted,

ASAI = 0.999479,

ASUI = 0.000521,

ENS   = 2736.7 kWhr/yr,

AENS = 114.03 kWhr/customer yr.

    The operating procedure utilized for a specific distribution configuration has an impact on the system reliability. In the case of the sample system shown in Figure 4.5, all the load points are normally supplied by the south end input supply. The second operating strategy which may be used is to supply eleven load points (1-5, 10-15) from the north end and the rest of the load points from the south end. The system therefore, operates as two

separate parts, each with its own normal input supply.
The two parts of the circuit are connected in the case of
an emergency in order to provide alternative supply to
the part in trouble by changing the connections inside
the switching cubicles IP1 or IP2. The comparison of the
reliability indices for the two operating strategies is
shown in Table 4.14. The input parameters used are the
same as those used for Table 4.13 except for the probabilities associated with the fuses and alternative supply. The probability of successful operation of a fuse
and the availability of alternative supply are assumed to
be 0.98 and 0.6 respectively for the studies shown in
Tables 4.14 and 4.15.

There is a considerable difference in the reliability
indices obtained for the two operating methods. In operating strategy 1, any failure of a cable section leads to
the interruption of all the load points in the system but
in the case of operating strategy 2, load points of only
one part of the circuit are interrupted.

The interruptions seen by the customer are due to
failures of either the components in the distribution
configuration or the main supply to the system. The effect of the latter is normally not included when evaluating the reliability indices of a distribution system.
Table 4.15 shows the impact of failures of the main supply on the reliability indices of the sample system shown
in Figure 4.5. The frequency and duration of input supply failures have been considered as 0.5 f/yr and 2 hrs
respectively.

Probability Distributions

The sample distribution system shown in Figure 4.5
has been analyzed using the computer program.
The following data were used in these studies:
failure rate of cable sections = 0.017 f/yr-km,
repair time of cable sections = 8 hr,

Table 4.14 - Reliability indices for operating strategies 1 and 2

(a) Basic indices

Operating strategy	Load point 3			Load point 12			Load point 18		
	λ f/yr	r hr	U hr/yr	λ f/yr	r hr	U hr/yr	λ f/yr	r hr	U hr/yr
1	0.6856	4.26	2.9204	0.6856	4.13	2.8318	0.6856	4.01	2.7482
2	0.3103	4.19	1.2989	0.3103	5.25	1.6293	0.3853	4.79	1.8473

(b) System indices

Operating strategy	SAIFI cus. int/ cus.	SAIDI cus. hr/ cus.	CAIDI cus. hr/ cus. int	ASAI	ASUI	ENS kWhr/ yr	AENS kWhr/ cus. yr
1	0.685	2.757	4.022	0.999685	0.000315	164.45	68.94
2	0.35093	1.630	4.645	0.999814	0.000186	978.0	40.75

Table 4.15 - Reliability indices including supply failures

(a) Basic indices

	Load point 3			Load point 12			Load point 18		
	λ f/yr	r hr	U hr/yr	λ f/yr	r hr	U hr/yr	λ f/yr	r hr	U hr/yr
Main supply fully reliable	0.6856	4.26	2.9204	0.6856	4.13	2.8318	0.6856	4.01	2.7482
Including failures of main supply	1.1856	3.56	4.2204	1.1856	3.48	4.1318	1.1856	3.41	4.0482

(b) System indices

	SAIFI cus.int/ cus.	SAIDI cus.hr/ cus.	CAIDI cus.hr/ cus.int	ASAI	ASUI	ENS kWhr/ yr	AENS kWhr/ cus.yr
Main supply fully reliable	0.6856	2.757	4.022	0.999685	0.000315	1654.5	68.94
Including failures of main supply	1.1856	4.057	3.422	0.999463	0.000463	2434.5	101.44

failure rate of transformers = 0.005 f/yr,
replacement time of transformers = 48 hr,
isolation and switching time = 3 hr,
probability of alternative supply availability = 1,
probability of fuses to clear the fault = 1.

The system has been studied for four different coef-
ficients of variation of the probability distributions
associated with the repair and restoration times. In all
the four cases, these times have been assumed to be normally distributed. Figures 4.6 to 4.11 give the results
for the system indices, SAIFI, SAIDI, ASUI, ASAI, ENS and
AENS respectively. The sum of ASAI and ASUI is equal to
1.0. The definition of percentile in Figure 4.9 is therefore, the probability that ASAI is equal to or greater
than the indicated value.

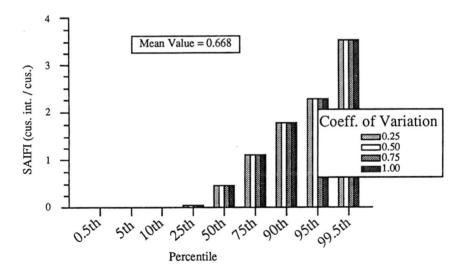

Figure 4.6 - Percentiles of SAIFI (repairs normally dis-
             tributed)

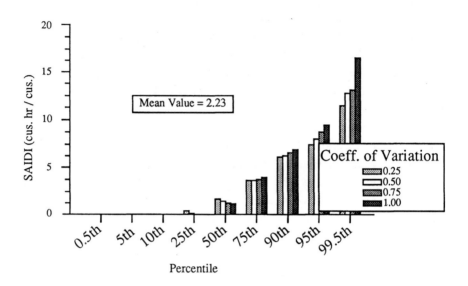

Figure 4.7 – Percentiles of SAIDI (repairs normally dis-
                  tributed)

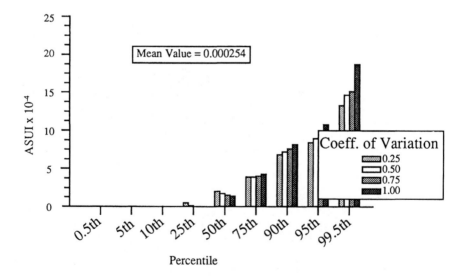

Figure 4.8 – Percentiles of ASUI (repairs normally dis-
                  tributed)

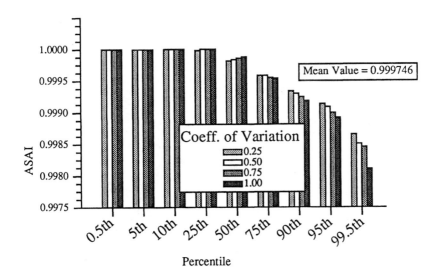

Figure 4.9 – Percentiles of ASAI (repairs normally distributed)

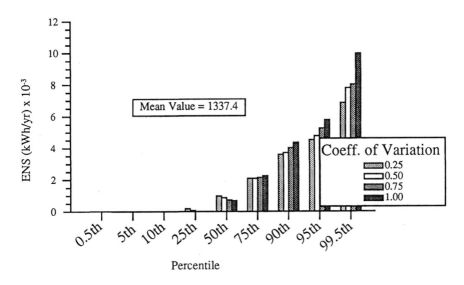

Figure 4.10 – Percentiles of ENS (repairs normally distributed)

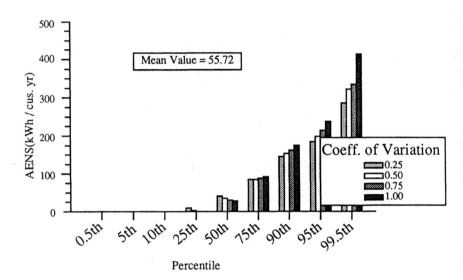

Figure 4.11 – Percentiles of AENS (repairs normally dis-
            tributed)

CONCLUSIONS

The basic techniques [1] to evaluate the average values of distribution system reliability indices can be ef-
fectively utilized to investigate the effect on the system performance of varying the component performance parameters, system restoration times, operating strategy,
system configuration etc.   These indices are relatively
easy to obtain and general computer programs have been
developed to calculate the individual load point indices
and the overall system indices.   The average values are
and will continue to be the main indices.   The distributions associated with these indices can, however, provide
an additional dimension which can provide a more physical
appreciation than the single point mean values.   Two ap-

proaches have been illustrated in this chapter to obtain the distributional information. The first approach is a general Monte Carlo simulation technique which theoretically can be used to analyze any configuration and use any known input information. The second technique is an analytical technique which permits direct evaluation of the average indices and the probability distributions associated with these indices. The direct analytical approach is a very practical and efficient technique which can be used in a wide range of system studies. Both techniques, however, can be used to generate the required distributional data. This information together with the basic average indices provides both a quantitative and a physical appreciation of distribution system reliability.

REFERENCES

1. Billinton, R. and Allan, R.N., "Reliability Evaluation Of Power Systems," Longman, London (England)/ Plenum Publishers, New York, (1984).
2. Billinton, R., Billinton, C.J. and Billinton, J.E., "Service Continuity Performance Of Canadian Electric Power Utility - A Historical Perspective," Proceedings of the 14th Inter-RAM Conference. IEEE 1987 CH. May 1987.
3. Patton, A.D., "Probability Distribution Of Transmission And Distribution Reliability Performance Indices," 1979 Reliability Conference for the Electric Power Industry, pp. 122-123, (1979).
4. Billinton, R., Wojczynski, E. and Rodych, V., "Probability Distributions Associated With Distribution System Reliability Indices," 1980 Reliability Conference for the Electric Power Industry, (1980).
5. Billinton, R. and Wojczynski, E., "Distributional Variation Of Distribution System Reliability Indices," IEEE Transactions PAS-104, No. 11, pp. 3152-3160, (1985).
6. Billinton, R., Wojczynski, E. and Godfrey, M., "Practical Calculations Of Distribution System Reliability Indices And Their Probability Distributions," CEA Transactions, (1981).
7. Billinton, R. and Goel, R., "An Analytical Approach To Evaluate Probability Distributions Associated With The Reliability Indices Of Electric Distribution System," IEEE 86 WM 156-4, (1986).
8. Johnsson, N.L., Nixon, E. and Amos, D.E., "Table Of

Percentage Points Of Pearson Curves," Biometrika,
Volume 50, pp. 459-498, (1963).

CHAPTER 5

ASSESSMENT OF RELIABILITY WORTH

INTRODUCTION

The function of a modern electric supply system is to provide electric power to customers at reasonable cost and acceptable levels of reliability. Reliability consists of both adequacy and security of supply; hence supply interruptions, regardless of the cause, constitute a reduction in reliability. Acceptable levels of reliability are normally assigned on the basis of experience with little or no attempt to directly relate target levels to economic or socioeconomic parameters. Increases in energy costs, construction costs and interest rates, the recognition of conservation and environmental concerns and the impacts of government and public groups have resulted in the need for a more rational and consistent approach to determining acceptable reliability levels. A major aspect of this approach is the attempt to assess the worth of power system reliability in order to be able to compare it with the costs of obtaining that reliability [1].

The basic concepts associated with reliability cost/ reliability worth evaluation were introduced in Chapter 1. These concepts are summarized in Figures 1.9 and 1.10. A major element in the evaluation of reliability worth is the customer cost associated with a loss of supply. The cost of interruption at a single customer load point is dependent entirely on the cost characteristics of that customer [2,3,4,5]. As the supply point in question moves away from the actual customer load point,

the consequences of an outage of the supply point involves an increasing number of customers. In the limit
as the supply point becomes the generating system i.e.
HLI, potentially all the system customers are involved.
The customer cost associated with a particular outage at
a specific point in the system involves an amalgamation
of the costs associated with the customers affected by
interruptions at that point in the system. This amalgamation or consolidation of costs is known as a composite
customer damage function (CCDF) [6].

The CCDF is an estimate of the cost associated with
power supply interruptions as a function of the interruption duration for the customer mix in the service area of
interest. Each customer or type of customer has a different cost for a particular outage duration and the
method for combining the individual costs is to perform a
weighted average according to the annual energy consumption of the individual customers or customer groups. Individual load point, sector, regional and system composite customer damage functions can be successively created
from their individual customer interruption cost and load
data.

It is important to realize that, while the evaluation
of power system reliability has become a well established
practice over the last decade, the assessment of the
worth of reliability or conversely the evaluation of
costs of losses which result from system unreliability is
still an immature technique. The major reason for this
is that quantification of interruption costs is a complex
and often subjective task.

A review of the literature reveals that the majority
of approaches used to assess reliability worth are based
on a determination of the impacts of interruptions, i.e.
the cost of unreliability [1]. In turn, the evaluation
of interruption impacts by means of customer surveys is
considered to yield relatively definitive results, and

such surveys are normally undertaken for each of the various user groups, e.g. commercial, industrial, residential, etc. [2,3,4,5]. The customer interruptions cost data presented in this chapter were obtained from a series of surveys conducted in 1980-81 by the University of Saskatchewan on behalf of the Canadian Electrical Association. The methodologies and concepts are universal. The data, however, are particular to Canada and as such may not be directly applicable to other regions or countries. All the costs given in this chapter are in 1980 Canadian dollars.

INTERRUPTION COSTS FOR COMMERCIAL CUSTOMERS

Scope Of Survey

In the mail survey of commercial users [3,4], which generated the data used in the analysis described in this section, respondents were asked to estimate the costs to their company for interruptions of various durations. The interruptions were to occur without warning on a Friday at 10:00 a.m. near the end of January. This was believed to be the worst time for the occurrence of an interruption. The surveyees were instructed to include lost business or sales, wages paid to staff who are unable to work, equipment or goods damaged, etc. in their estimates, but not to include sales or business that could be made after the interruption ceases. The questionnaire listed factors contributing to the cost and respondents were asked to indicate, for various interruption scenarios, the component costs that made up their estimates. This form of question was intended to aid respondents in making estimates and to obtain some information on the relative severity of the different components of effects. These interruption cost estimates were then analyzed as a function of whether the user had a standby system or not, and if so, what type of standby system.

Table 5.1 lists the contributing factors used in the

questionnaire. Table 5.2 indicates the composition of the
respondent sample by standby condition. Although not
shown in this listing, it was observed that the largest
users of standby equipment were retail services followed
by food stores and retail trade.

Table 5.1 — Cost factors used in commercial cost estimates

Paid staff unable to work
Loss of sales
Start up costs
Spoilage of food
Damage to equipment/supplies
Other costs or effects

Table 5.2 — Breakdown of respondents into categories

No standby equipment	87%
Battery standby system	10%
Engine-driven system	4%
Other system	1%
Number of respondents	979

Note: The percentages shown do not total 100 because
some users had more than one type of standby system.

The interruption scenario, namely Friday at 10:00
a.m. near the end of January, was considered to be the
"base case" for cost estimation. The variation in interruption cost relative to this base case with month of
year, day of the week and time of the day was obtained in
the questionnaire using a tabular format [4]. Using
Scaleogram analysis, the relative costs for each time
frame were ordered. Although this ordering does not place
months in chronological sequence for any of the three
standby categories, and although the ordering for all
three categories is not identical, the following consistent seasonal groupings can be made:

Summer months — May through August inclusive,

Fall & spring months — April, September, October, November,

Winter months - December through March inclusive.

The purpose of such grouping is to describe seasonal variations rather than attempt to describe or quantify monthly variations. The variation of cost estimates with season can be found using the data provided by the respondents and grouping these according to the "seasons" identified above. There were no monthly cost estimates above the "25% higher than January cost" reported, and only a few respondents (well below 1%) indicated costs below the "25% less than January cost" category.

Costs Of Interruptions

The base case cost estimate and the monthly variation of cost as provided by the respondents were used to determine an interruption cost estimate for each of the seasonal groups identified above. Respondent's annual energy consumption (kWh) and annual peak demand (kW) were obtained from the utilities (if the respondent had given permission to do so). Each respondent's cost estimates were normalized with respect to annual peak demand ($/kW) and annual energy consumption ($/kWh). It should be noted that this normalization procedure does not yield cost estimates for energy or demand not supplied, which would be highly desirable. However, to determine such values would require additional data, namely the demand at the time of interruption and its variation during the duration of the interruption. Such information is not available, and the normalization approach used in this work is consistent with that used by others [7,8]. The average dollars per interruption and the normalized cost estimates for the no standby system and engine system categories for the three seasonal groups identified earlier were calculated and are presented in Figure 5.1.

These cost estimates were calculated by weighting the cost estimates for each Standard Industrial Classifica-

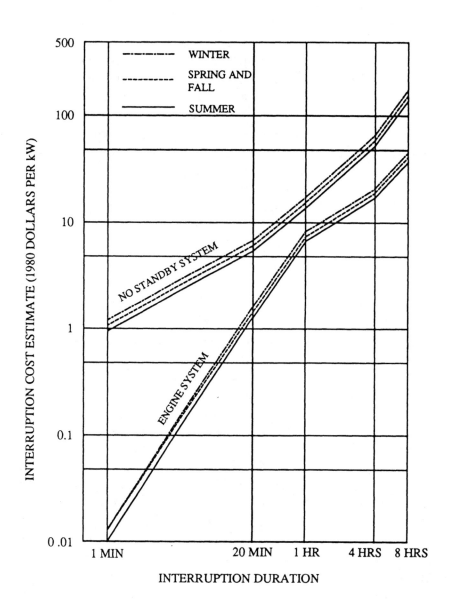

Figure 5.1 – Seasonal variation of commercial customers'
           interruption costs grouped by standby
           category

tion (SIC) group category according to the relative
amount of energy consumed by users of that category in
the Manitoba Hydro service area. The responses within
each category were aggregated either by annual energy
consumption (for $/kWh estimates) or by annual peak de-
mand (for $/kW estimates). Thus, aggregated weighting
was achieved by summing the dollar costs for all respondents in each SIC group category and dividing this total
cost by the total of the energy consumptions (or peak demands) for those same respondents, and then multiplying
this ratio by the decimal fraction of the total service
area energy which is consumed by that SIC group. Only
respondents for which the cost estimates and the energy
consumption and demand figures were available were included in the calculations.

Figure 5.2 illustrates the variation of $/kW cost es-
timates as a function of interruption duration for the
three system categories. It is observed that the outage
costs for users having standby systems are less than
those with no standby. It must be recognized that in
situations of this nature, categorizing a given popula-
tion according to some variable (standby category in this
case) and determining some quantity ($/kW in this case)
does not suggest that observed variations of the determined quantity are assignable to the categorization variable. In fact, other variables may well correlate with
and be the cause of the observed variation. In this case,
for example, the lower losses reported by users with
standby is perhaps largely due to size of the user's
enterprise, i.e. larger users are more likely to afford
standby systems and the normalization procedure of dividing by a larger number results in lower normalized cost
estimates. However, since the other variables are unknown, the variable of interest, namely standby category,
is an acceptable categorization criterion. Figure 5.1
shows the seasonal variation for the no standby and en-

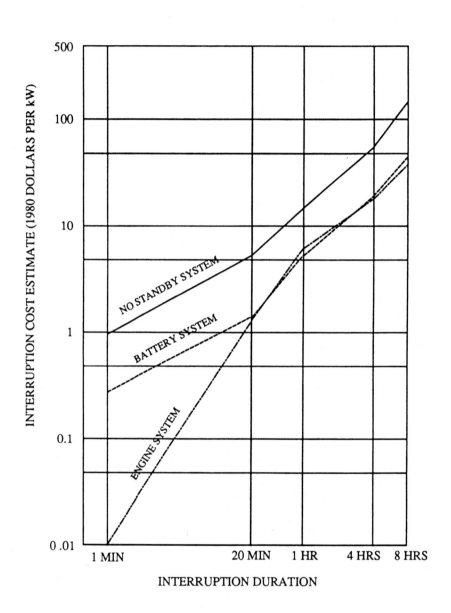

Figure 5.2 – Commercial customers' interruption costs
          grouped by standby category

gine-driven standby categories for the same situation as
in Figure 5.2. It is obvious that the seasonal variation
is insignificant in comparison with standby category and
in an absolute sense. It seems apparent that, except for
peculiar situations, seasonal variations of costs need
not be considered.

Customer Damage Function

The results shown in Figure 5.2 illustrate that the
commercial outage costs are sensitive to standby system
availability. Not evident in the figure is that there is
a large variation in the costs among the various customer
categories, although the variation is less within an SIC
group than between them [4]. Therefore, the expected accuracy of the derived customer damage function [11] will
improve as the representation of the customer mix in
terms of customer type, size, standby category, etc. more
nearly represents the service area in question. As an
example, consider the Manitoba Hydro load mix with 10% of
its energy consumed by the commercial sector. This is
further subdivided into the three standby subgroups.
Table 5.3 lists the average values of consumption and
peak demand for the respondents in each subgroup. Table
5.4 presents three possible mixes of customers according
to standby category with mix #1 representing (approximately) Manitoba Hydro's situation. Aggregated customer

Table 5.3 – Respondents annual energy consumption and
          peak demand

	Respondents with		
	No standby system	Battery system	Engine system
Average energy consumption (kWh)	101,761	284,230	991,184
Average peak demand (kW)	47	86	314

Table 5.4 – Commercial load mixes (by % of energy con-
          sumption)

	User Mix		
Respondents with	#1	#2	#3
No standby system	1	1	2
Battery standby system	2	4	2
Engine standby system	7	5	6

costs for the commercial sector are obtained by weighting
the user group outage costs by percentages of their load
mixes and summing these costs.   The results are shown
graphically in Figure 5.3 and Figure 5.4.   Figure 5.3

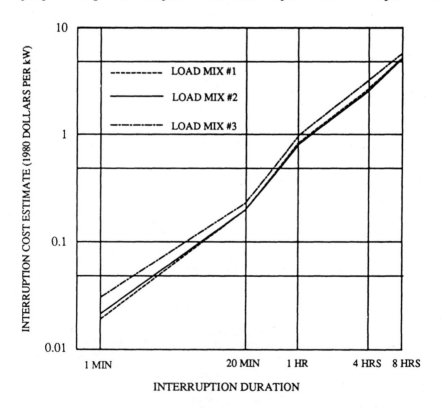

Figure 5.3 – Commercial customers' interruption costs in
             winter for three load mixes by standby cat-
             egory

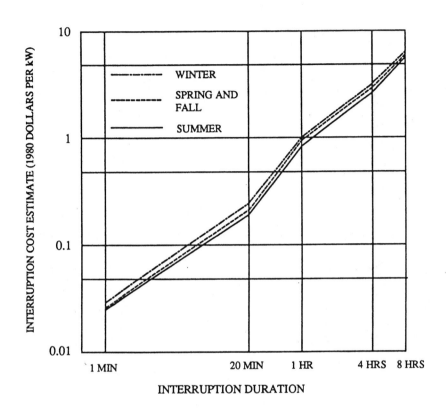

Figure 5.4 - Seasonal variation of commercial customers'
           interruption costs for load mix #3

illustrates the winter costs for the three load mixes,
while Figure 5.4 shows the seasonal effect on cost
estimates for load mix #3. The effect of determining aggregated costs weighted by standby category as compared
with gross aggregated costs (irrespective of standby category) is presented in Table 5.5 and Figure 5.5. These
data suggest that the relatively large variations of cost
with customer type and type of standby tend to be diluted
during the aggregating process.

Table 5.5 – Aggregated costs weighted by standby
            category compared with gross aggregated
            costs for mix #1

Interruption duration	Aggreg. costs weighted by standby category	Gross aggreg. cost
1 min	0.017	0.028
20 min	0.19	0.21
1 hr	0.75	0.59
4 hr	2.39	2.15
8 hr	5.55	6.31

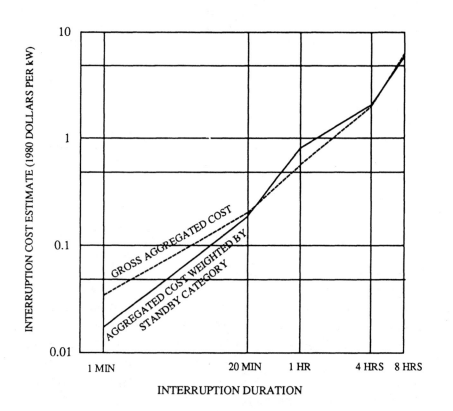

Figure 5.5 – Aggregated commercial cost estimates ob-
             tained by weighting according to standby
             category compared with estimates obtained by
             simple gross aggregation

Conclusions

This section has presented a summary of studies conducted to determine the effect on outage costs, and
thereby on the customer damage function, of commercial
consumers having electric standby systems. Observations
and conclusions reached can be summarized as follows.

Consumption and demand normalized outage costs for
users having no standby systems are higher (typically by
3 times) when compared with users having standby systems.
No significant differences were observed between battery
and engine-driven standby systems. The variation of outage costs with time of year tend to have a negligible effect on the weighted, aggregated costs which in turn are
used to establish a commercial sector or composite customer damage function.

The commercial component of a composite customer damage function can be determined given the load composition
as to customer category and standby availability and
type. Although there tend to be significant variations
with type of customer and standby category and lesser
variation with season, the aggregated customer cost for
the commercial component tends to be relatively insensitive to variations of these variables. Furthermore, the
commercial sector often represents a small fraction of
the total electrical load (e.g. 10%). Therefore, it
should be possible to generate a representative commercial customer damage function for a given service area
which will be applicable for a considerable range of load
mix.

INTERRUPTION COSTS FOR INDUSTRIAL CUSTOMERS

Scope Of Survey

This section discusses the effect and variation of a
number of variables germaine to the small industrial component of a composite customer damage function [4]. In

particular, the effect of having standby systems is considered since it was previously noted that respondents
having standby generating capacity have lower demand- or
consumption-normalized outage costs.   A comprehensive
study [3] was made to find the outage costs for the respondents with and without standby systems, and the variation of the outage costs with time of the year, week and
day. The outage costs were then used to create a localized customer damage function for the industrial category
alone, and to study its sensitivity to load mix and the
time of the year.   The data base used for this study
originated from the customer surveys reported in Reference 3.

In the survey of small industrial users [4], respondents were asked to estimate the costs to their company
for interruptions of various durations.   The interruptions were to occur without warning on a Friday at 10:00
a.m. near the end of January. Industrial surveyees were
told to include in the estimates, plant and equipment
damage, raw materials and finished product spoilage or
damage and the cost of special procedures to restart production (e.g. extra clean up, maintenance, checkups,
etc.).   Production lost during the failure and restart
time was to be evaluated as the estimated revenue of
product not made, less the expenses saved in labor materials, utilities, etc. If production could be made up
later during slack time or overtime, that portion was not
to be included. Other costs such as the cost of operating standby equipment or of special procedures to prevent
damage could be listed as well. A set of contributing
factors was provided, and respondents were asked to estimate the cost associated with each of the contributing
factors for various interruption scenarios. Addition of
these component costs gives the total cost estimate for
each customer and for each interruption scenario. These
interruption cost estimates were then analyzed as a func-

tion of whether the user had a standby system or not, and if so, what type of standby.

Table 5.6 lists the contributing factors used in the questionnaire. Table 5.7 indicates the make up of the sample by standby condition.

Table 5.6 – Cost factors used in small industrial cost estimates

1. Plant/equipment damage
2. Raw material/finished product damage
3. Start up costs (extra cleanup or maintenance)
4. Production loss (during failure and restart time)
5. Overtime to make up lost production
6. Other costs or effects (e.g. operating standby equipment)

Table 5.7 – Breakdown of respondents into categories

No standby equipment	87%
Battery standby system	6%
Engine-driven system	8%
Other system	1%
Number of respondents	416

Note: The percentages shown do not total 100 because some users had more than one type of standby system.

A similar analysis to that described previously for commercial users was done. This produced slightly different seasonal groupings. These are:

Winter months – December through March inclusive,

Summer months – June through September;,

Spring and fall months – April, May, October and November.

The "base case" estimate, namely Friday at 10:00 a.m. near the end of January, was then used to obtain an interruption cost estimate for each of the above seasonal groupings. In order to calculate the outage cost variation with season, further analysis was undertaken to find the breakdown of respondents indicating variation in cost estimates with season. It should be noted that no re-

spondents indicated more variation than 25% above or below the January cost.

Costs Of Interruptions

Respondent's annual energy consumption (kWhr) and annual peak demand (kW) were obtained from the utilities
(if the respondent had given permission to do so). For
customers without demand readings, the annual peak demand
was estimated using approximate load factor values for
each SIC. Demand values provided were peak kVA not kW.
As no reliable power factor data were available and because the resulting error was deemed acceptable, the kVA
values were assumed to be in kW. The cost estimates and
aggregate weighting were evaluated using the same technique described previously for commercial users.

The variation in $/kW with interruption duration for
the three standby categories and the summer season is
shown in Figure 5.6. Figure 5.7 shows the interruption
cost estimates for the three seasons for the no-standby
and battery standby system categories.

Customer Damage Function

The results shown in Figure 5.7 illustrate that the
outage costs are sensitive to standby system generating
capability. Furthermore, it has been observed that the
variability in the costs is high for virtually all customer categories. Hence, it is necessary to create a local customer damage function for the industrial sector
reflecting both customer mix and standby situations [12].
The calculated local customer damage function can then be
directly substituted in the calculation of any larger
area customer damage function.

Industrial respondents consume on average 20% of the
energy generated. Creation of a load mix, for the three
sets of respondents is necessary for the calculation of
the customer damage function. As an example, consider

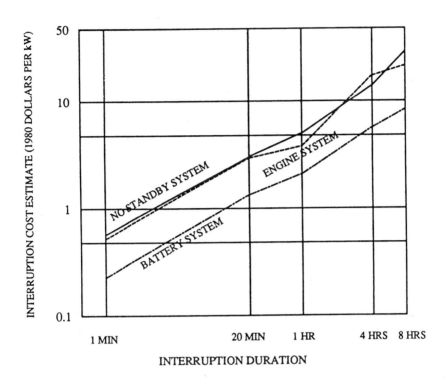

Figure 5.6 – Interruption cost estimates for the three
            categories of respondents for summer

the Manitoba Hydro load mix with 19% of its energy consumed by the industrial sector. This is further subdivided into the three subgroups, namely respondents with
no standby systems, respondents with battery standby systems and respondents with engine-drive standby systems.
Table 5.8 lists the average values of consumption and
peak demand for respondents in each subgroup. Table 5.9
lists three possible mixes of customers according to
standby grouping, with Mix #1 representing (approximately) Manitoba Hydro's situation. Aggregated customer costs
for the industrial sector are obtained by weighting the
user group outage costs by percentages of their load
mixes and summing these costs for each duration. The

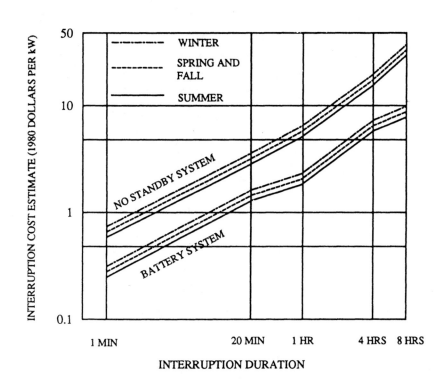

Figure 5.7 – Interruption cost estimates for the three
            seasons for respondents of the no–standby
            and battery standby system categories

Table 5.8 – Respondents annual energy consumption and
            peak demand

| | Respondents with | | |
	No standby system	Battery system	Engine system
Average energy consumption (kWh)	1,138,301	45,912,045	31,994,750
Average peak demand (kW)	351	7,765	6,239

Table 5.9 — Industrial load mixes (by % of energy con-
           sumption)

| | User mix | | |
Respondents with	#1	#2	#3
No standby system	1	2	3
Battery standby system	11	9	8
Engine standby system	7	8	8

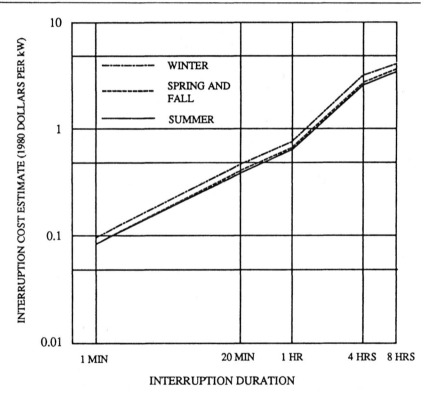

Figure 5.8 — Customer damage function for different sea-
             sons for load mix #3

results are illustrated in Figures 5.8 and 5.9.

The local customer damage function (CDF) for the load
Mix #1 which is approximately the Manitoba Hydro load mix
was compared with the previously used industrial contri-

bution to the composite customer damage function. Table
5.10 shows the results of this comparison, with Figure
5.10 graphically illustrating the difference.

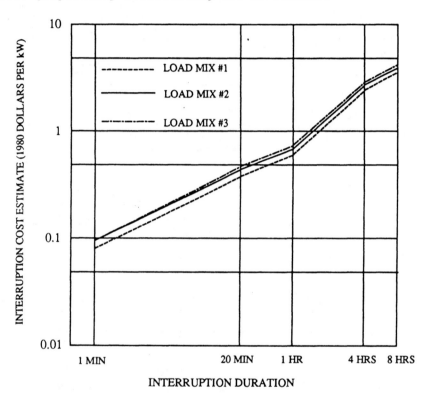

Figure 5.9 – Customer damage function for winter

Table 5.10 – Gross aggregated, CDF with aggregated CDF
            weighted by type of standby system for Mix
            #1 (in 1980 $/kW)

Interruption duration	Mix #1 Aggreg. CDF weighted by standby category	Gross aggreg. CDF
1 min	0.08	0.13
20 min	0.40	0.55
1 hr	0.61	0.99
4 hr	2.27	2.64
8 hr	3.27	5.24

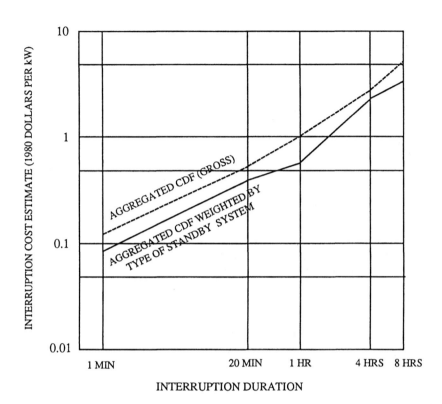

Figure 5.10 - Comparison of gross aggregated CDF with ag-
             gregated CDF weighted by type of standby
             system - industrial sector

Respondents having standby systems estimate the costs
associated with plant/equipment damage start up costs and
production losses to be much higher compared to respondents having no standby system. Respondents having no
standby systems estimate the lowest cost for virtually
all the factors. In the estimation of cost associated
with overtime paid to make up lost production, respondents having battery systems have indicated higher costs
compared to the other two categories which are closely
matched. This is supported by the fact that 35% of the
respondents having engine-driven systems use it mainly to
maintain production thereby not having to pay a great

deal of overtime or hire extra employees, whereas respondents having battery systems do not use their system at
all to maintain production, and also employ 240% more
part time employees compared to respondents having engine-driven systems.

The only factor where costs estimated by a category
other than the battery system standby group is higher is
associated with other costs or effects. In this case respondents having engine-driven systems estimate a higher
cost. A partial explanation of this observation may be,
that costs associated with operating engine driven standby equipment is higher than the costs involved in operating battery standby equipment.

Conclusions

This section has summarized studies conducted to determine the effect on outage costs for respondents having
standby system capacity and thereby on the customer damage function. Conclusions reached can be summarized as
follows.

Costs of interruptions for the battery system are the
lowest when compared with the engine systems and no
standby system categories. It is possible that engine
drive systems have higher outage costs compared with battery systems because of the cost of operating the system.
This is supported by the fact that the factor "other
costs and effects", which includes the cost of operating
standby systems, is highest for engine systems.

Interruption costs for the various seasons have been
calculated and studied. The differences in outage costs
are not very significant and hence it can be concluded
that outage costs calculated for winter can be used as an
upper bound for other seasons. The difference in outage
costs for winter weekdays and weekends is significant
enough to conclude that outage costs for weekends in summer could probably have the lowest outage costs and may

be significant enough to affect the customer damage function when compared to outage costs in winter.

Industrial respondents consume approximately one-
fifth of the total energy generated. The industrial customer damage function is therefore a significant component in a composite customer damage function. The results show that there is no significant difference between the industrial customer damage functions for the
three load mixes considered in this study. It appears
reasonable therefore that a representative industrial
customer damage function can be obtained for a wide range
of studies.

INTERRUPTION COSTS FOR RESIDENTIAL CUSTOMERS

Scope Of Survey

The data used in this study are based on 4665 usable
responses from a residential survey resulting from a
mailing sample of 13,359 questionnaires [3,5]. The survey involved eight Canadian utility service areas located
in the Maritimes, the Prairie Provinces and interior
British Columbia. Respondents were asked to provide interruption cost estimates based on their anticipated preparatory actions and also based on their opinion of expected rate changes for given changes in reliability.
The breakdown of respondents by dwelling type and their
average annual energy consumption are provided in Table
5.11.

Table 5.11 - Respondent proportions and consumption by
            dwelling type

Dwelling type	% of respondents	Consumption (kWh)
Single house	86.0%	11,960
Apartment	9.0%	6,527
Mobile home	3.0%	14,320
Suite in a house	2.0%	7,308

Total number of respondents = 4665

Costs Of Interruptions

Effect Of Preparatory Actions.   Respondents were
asked to suppose that they have been told by their power
company that power failures will occur after 4:00 p.m. on
winter weekdays, but that the exact days or times are not
known.   They were then asked to predict which actions
their household might take in preparation for the failures.   Respondents were asked to choose one or more of
the following six actions: make no preparations, purchase and use a candle at $0.25 per hr, an emergency lan-
tern at $0.50 per hr, an emergency stove at $1.50 per hr,
purchase or rent and use a small generator for lighting
and small appliances at $5.00 per hr, or a larger gen-
erator for the full household load at $10.00 per hr.
These costs were used to compute an estimate of the cost
of preparations that respondents indicate they are willing to undertake to reduce or eliminate the adverse effects of the interruption; this is called the preparatory
action cost estimate.   In addition to reporting costs on
a cost/interruption basis, the costs were normalized with
respect to each respondent's actual annual consumption
($/MWh) and estimated peak demand ($/kW) for comparison
and planning purposes.   The estimated peak demand was obtained from users' annual consumption history by assuming
a load factor of 23%.   Table 5.12 lists the average $/in-
terruption, $/MWh and $/kW estimates determined from preparatory actions for each of the various dwelling types
under consideration.   All estimates are presented on a
per interruption basis to make them consistent.   The $/kW
values for the 1 min and the 8 hr interruption were estimated by extrapolation.   This extension is necessary when
these cost estimates are combined with the costs of other
sectors in establishing a global customer damage function.   Figure 5.11 graphically illustrates these data.

Effect Of Rate Changes. To provide a proper basis
for answering the rate change questions, respondents were
first asked to estimate their monthly electricity bill.
They were also asked their preference about wanting an
increase in rates with a corresponding decrease in the
number of failures, a decrease in rates with an increase
in the number of failures or for the situation to remain
unchanged. The results are given in Table 5.13.

Table 5.12 - Average preparatory action interruption cost
             estimates for winter weekdays after 4:00
             p.m.

Interruption Scenario	Single house	Apt.	Mobile home	Suite in a house
Cost per interruption ($)				
20 min monthly	0.22	0.20	0.20	0.25
1 hr monthly	1.19	0.93	1.43	0.84
4 hr monthly	12.00	9.00	14.61	12.24
4 hr weekly	18.58	14.63	21.41	18.41
Cost per interruption divided by annual energy consumption ($/MWh)				
20 min monthly	0.031	0.047	0.036	0.054
1 hr monthly	0.162	0.276	0.243	0.222
4 hr monthly	1.656	2.480	1.686	1.975
4 hr weekly	2.970	3.953	2.787	3.076
Cost per interruption divided by annual peak demand ($/kW)				
1 min monthly	0.00*	0.00*	0.00*	0.00*
20 min monthly	0.06	0.10	0.07	0.11
1 hr monthly	0.33	0.56	0.49	0.45
4 hr monthly	3.34	5.00	3.40	3.98
4 hr weekly	5.98	7.96	5.62	6.20
8 hr monthly	10.00*	15.00*	8.90*	12.50*

* estimated by extrapolation

The first rate change question instructs respondents
to suppose that the normal electric system has become
subject to more frequent power failures. They are asked
whether they would choose an alternative assured electri-

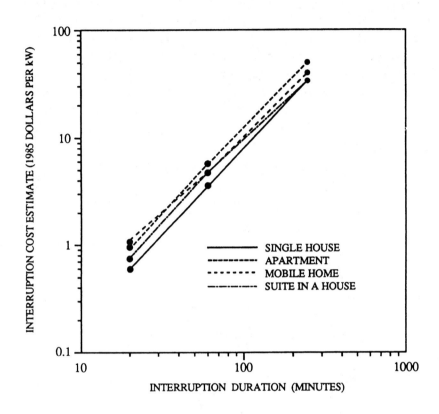

Figure 5.11 - Interruption cost estimates - 1980 $/kW

Table 5.13 - Respondent's reliability preference

	Single house	Apt.	Mobile home	Suite in a house
Preference	% of Respondent's indicating			
Rate increase, failures decrease	7	10	8	6
Rate decrease, failures increase	7	8	9	6
No change	86	82	83	88

cal power supply, i.e. one without any failures, at various premiums above the cost of the normal system.   The
average value of the maximum amount per month that respondents are willing to pay for an assured system is
calculated.   The values are normalized with respect to
consumption and with respect to peak demand for comparison purposes and are presented in Table 5.14.

Table 5.14 – Rate increase interruption cost estimates

Average maximum amount above the cost of the normal supply that customers are willing to pay for the assured
supply.

Interruption Scenario	Single house	Apt.	Mobile home	Suite in a house
Cost per interruption ($)				
4 hr monthly	5.31	5.09	6.08	3.79
4 hr weekly	8.15	7.50	8.25	5.84
1 hr daily	8.17	7.50	7.97	6.78
Cost per interruption divided by annual energy consumption ($/MWh)				
4 hr monthly	0.68	1.23	0.684	1.10
4 hr weekly	1.04	1.85	1.01	1.64
1 hr daily	1.04	1.72	0.95	1.59
Cost per interruption divided by annual peak demand ($/kW)				
4 hr monthly	1.37	2.48	1.38	2.22
4 hr weekly	2.10	3.73	2.03	3.31
1 hr daily	2.10	3.46	1.91	3.21

The second rate change question is complementary to
the willingness-to-pay approach described above, namely
to determine a willingness to accept a decrease in reliability for a decrease in rates. Respondents were asked
to indicate the minimum percentage decrease in their
rates for them to choose a specific reduced reliability.
The average value of the minimum rate decrease for respondents to choose the option of a cut in rates together

with an increase in the number of failures to a 4 hr
failure occurring once a month is calculated. These values are also normalized with respect to consumption and
peak demand and are shown in Table 5.15.

Table 5.15 - Rate decrease interruption cost estimates

Average minimum rate reduction customers will accept for
a given reduction in supply reliability.

Interruption scenario	Single house	Apt.	Mobile home	Suite in a house
Dollars per interruption ($)				
4 hr monthly	11.32	7.39	13.04	6.75
Cost per interruption divided by annual energy consumption ($/MWh)				
4 hr monthly	1.29	1.52	1.49	1.24
Cost per interruption divided by annual peak demand ($/kW)				
4 hr monthly	2.59	3.06	2.99	2.50

Discussion. It is important to note that the cost
estimates obtained from the three approaches are not
strictly comparable, the major reason being that the scenarios for these approaches are not identical. The preparatory action estimates are for a scenario in which
failures occur after 4:00 p.m. on winter weekdays. The
rate increase and decrease estimates are more general and
concern failures in the day or evening and were not nec-
essarily restricted to winter. These were made more general in order that the responses would more clearly relate to the actual situation faced by the users.

Table 5.14 indicates that, on a per interruption basis, the rate increase estimate for daily failures is
similar to or less than the estimate for weekly failures.
This indicates the presence of a non-linear saturation
process when respondents are considering possible rate

increases, i.e. the respondents' indicated willingness-
to-pay appears to vary non-linearly with interruption
frequency. This saturation may occur because respondents
perceive the total monthly amount as being too large a
portion of their income. There is no conclusive evidence
or explanation as to why this non-linear variation occurs. This saturation effect does indicate, however, the
uncertainty inherent in obtaining an estimate with the
scenario of very frequent failures and applying it in a
situation of less frequent failures.

Analysis By Sex. A significant attribute of the impact of interruptions to residential users is its intan-
gible nature: the "value" of the "loss" due to inconve-
nience or disruption of housekeeping or leisure activities involves the perception of the individual(s) affected. Therefore an analysis by respondent sex was conducted to determine if there was any correlation.

The analysis performed was similar to that conducted
for dwelling type. Interruption cost estimates were
grouped into three categories of respondents namely;
male, female and joint response. The breakdown of respondents by sex is provided in Table 5.16. No significant difference was found to exist in the cost estimates
among the three categories.

Table 5.16 - Respondent's sex

Respondent sex	% of respondent's
Male	47
Female	22
Joint response	30
Number of respondents	4665

Transferability Of Costs. An important aspect concerning the application of cost estimates derived from
customer surveys is the question of transferability from
one geographical region to another. An attempt was made

to answer this question by analyzing results from the
various geographical regions considered in the survey.

The survey covered eight different utility service
areas. Factors included in the questionnaire which are
related to the issue of transferability were investigat-
ed, including number of failures, number of disruptive
failures, number of 4 hr failures, rating of undesirable
effects, interruption cost estimates, electrical heat us-
age, electrical consumption, etc. It was found that residential cost estimates obtained for one geographical
region may not be valid for another region.

Customer Damage Function

The results previously shown illustrate that outage
costs are a function of dwelling type. Therefore, the
contribution of the residential sector to the composite
customer damage function [10] should be the addition of
the individual dwelling type outage costs weighted in
proportion to their energy utilization. In other words,
a localized residential sector customer damage function
may be calculated using the local load mix by dwelling
type. This function in turn could be used in the determination of an overall composite customer damage function. In the determination of a residential customer
damage function, the $/kW values obtained from the action
cost estimates were used because it is believed that
these most accurately reflect residential customer losses
[3]. The contributions of the various dwelling types to
the load mix were computed based on the number of occu-
pied dwellings according to Statistics Canada data and
the average annual consumption for each customer type.
Table 5.17 gives the approximate contributions of the
various dwelling types to the residential load mix on a
Canada-wide basis.

The cost estimates for the various dwelling types are
proportionally weighted to their respective energy con-

Table 5.17 - Approximate contribution to residential load
           mix

Dwelling type	% contribution to load mix
Single house	80
Apartment	16
Mobile home	3
Suite in a house	1

sumptions within each service area. These weighted costs
are summed for each interruption duration to provide the
total cost for the area for that duration. Customer load
mix #1 shown in Table 5.18 is Manitoba Hydro's approximate load mix. The total residential consumption in that
service area represents 31% of the total energy utilization. The second load mix is the approximate mix for
Ontario Hydro's Central Region.

Table 5.18 - Load mixes by % of energy consumption

Dwelling type	Mix #1	Mix #2
Single house	24	31
Apartment	5	6
Mobile home	1	1
Suite in a house	1	1

The customer damage function is calculated by weighting the outage costs shown in Table 5.12 by the user load
mixes shown in Table 5.18, and totalled for each duration. The resulting aggregated costs are shown in Table
5.19.

The customer damage function for load mix #1 is compared with one calculated for the same situation but not
weighted by dwelling type (as previously reported and
used as the residential contribution in determining a
composite function [6]). Table 5.20 shows the results of
this comparison.

A major observation from these results is that the
residential component of the customer damage function is

Table 5.19 – Aggregated residential customer cost (1980
            $/kW) weighted by dwelling type

For winter weekdays after 4:00 p.m.

Interruption duration	Mix #1	Mix #2
1 min	0.0	0.0
20 min	0.02	0.03
1 hr	0.12	0.15
4 hr	1.13	1.41
8 hr	3.36	4.21

Table 5.20 – Comparison of the residential sector damage
            function calculated with and without weight-
            ing by dwelling type (for load mix #1)

For winter weekdays after 4:00 p.m.

Interruption duration	Weighted by dwelling type	Gross aggreg. (not weighted)
1 min	0.00*	0.00*
20 min	0.02	0.02
1 hr	0.12	0.10
4 hr	1.13	0.98
8 hr	3.36*	3.10*

* Estimated by extrapolation.

not highly dependent on load mix. This is mainly because
the load mix is very heavily biased towards the single
house dwelling type. It should be noted however, that a
bias towards the single house dwelling type may be present because the load mix is general, i.e. all Canada. It
is entirely possible that a vastly different breakdown by
dwelling type may exist within a particular service area
in which case dwelling type may become an important secondary factor.

To test this hypothesis, a hypothetical load mix
based on the contribution of various dwelling types as
shown in Table 5.21 was used to generate the load mixes
shown in Table 5.22. The resulting composite customer
damage function based on this hypothetical load mix is
shown in Table 5.23.

Table 5.21 - Contribution to residential load mix (hypo-
            thetical case)

Dwelling type	% contribution to load mix
Single house	60
Apartment	28
Mobile home	8
Suite in a house	4

Table 5.22 - Load mixes by % of energy consumption (hypo-
            thetical case)

Dwelling type	Mix #1	Mix #2
Single house	19	23
Apartment	9	11
Mobile home	2	3
Suite in a house	1	2

Table 5.23 - Residential sector customer damage function
            (for hypothetical case)

Interruption duration	Mix #1	Mix #2
1 min	0.00	0.00
20 min	0.02	0.03
1 hr	0.13	0.16
4 hr	1.19	1.50
8 hr	3.55	4.47

Comparison of Tables 5.14 and 5.23 shows that even a significant change in the contribution of various dwelling types does not seriously affect the residential sector damage function. Therefore, for all practical purposes, dwelling type may be disregarded as a secondary factor in the make up of the residential sector customer damage function.

The impact of using transferred interruption costs in the determination of a customer damage function was also studied. This was accomplished by assuming that Hydro Quebec does not have interruption costs for its service area, and that Winnipeg Hydro's interruption cost values are to be used to determine a customer damage function for Hydro Quebec. It is also assumed that 39% of the to-

tal energy generated by Hydro Quebec is used by the residential sector. The 39% component is hypothetical but
reasonable as residential consumers in this service area
have among the highest average annual energy consumption
of all the service areas studied. A second hypothetical
case with a load component of 31% is also considered.
Note that the 39% and 31% residential components are the
same fractions used in an earlier example, Mix #2 and Mix
#1 respectively.

The actual cost estimates obtained from the Hydro
Quebec and Winnipeg Hydro service areas are shown in
Table 5.24. The Hydro Quebec customer damage function is
calculated for each of the two mixes in each of two possible ways, namely, using Hydro Quebec's own data and applying Winnipeg Hydro's data to the Quebec (assumed)
situation. The resulting customer damage functions are
presented in Table 5.25. The results in Table 5.25 indi

Table 5.24 - Interruption cost estimates for Hydro Quebec
             and Winnipeg Hydro

Interruption duration	$/kW for Hydro Quebec	$/kW for Winnipeg Hydro
1 min	0.00	0.00
20 min	0.04	0.16
1 hr	0.24	0.62
4 hr	2.52	6.04
8 hr	8.00	19.50

Table 5.25 - Sector customer damage function for Hydro
             Quebec (1980 $/kW)

Interruption	Using Hydro Quebec data		Using Winnipeg Hydro data	
	Mix #1	Mix #2	Mix #1	Mix #2
1 min	0.00	0.00	0.00	0.00
20 min	0.01	0.02	0.05	0.06
1 hr	0.07	0.09	0.19	0.24
4 hr	0.78	0.98	1.84	2.36
8 hr	2.48	3.12	6.05	7.61

cate a major change in the customer damage function due
to the data used. These results suggest that transferability of interruption costs is questionable for the
residential sector.

Conclusions

This section presents a brief summary of the studies
conducted to determine estimates of the perceived costs
incurred by the various residential categories (single
house, apartments, mobile homes and suite in a house) due
to electric service interruptions. Preparatory action
cost estimates indicate that apartment dwellers have the
highest outage costs and mobile homes the lowest. A
saturation effect as the severity of the interruption
scenarios increases was observed in responses to the rate
change questions. This clearly indicates the uncertainty
inherent in obtaining estimates with scenarios of very
frequent failures and applying these to a situation of
less frequent failures.

Analysis of cost estimates by respondent sex did not
reveal any serious differences. Therefore, for all practical purposes, it can be concluded that respondent sex
does not affect the residential contribution to the composite customer damage function. Although significant
variations with type of dwelling are observed, the customer damage function reflecting the residential sector
tends to be relatively insensitive to variations in load
mix.

Transferability of residential cost estimates from
one geographical region to another and its effect on the
customer damage function was also studied. The results
indicate that residential cost estimates obtained in one
geographical region may not be valid in the determination
of a customer damage function for another region. The
potential limitations of the survey results are as
follows:

(a) the results concern respondent predictions of their
    actions in response to interruptions and the expected
    costs and effects arising from various interruption
    scenarios; their actual actions and cost may differ
    from their predictions,
(b) respondents may not have experienced some of the sce-
    narios,
(c) the estimates of interruption cost pertain to specif-
    ic interruption scenarios. While indications of the
    variation in cost with factors such as time of occur-
    rence were obtained, dollar cost estimates could not
    be obtained for every possible scenario,
(d) interruption cost is likely to vary with the level of
    reliability. The estimates assume a reliability level
    not too different from that presently experienced.

INTERRUPTION ENERGY ASSESSMENT RATE

Background

Composite customer damage functions (CCDF) can be
utilized [13] at each of the three hierarchical levels to
give an assessment of reliability worth. At HLIII, the
CCDF may involved a single customer or customer type. At
HLII, the load point of the composite system may serve a
wide range of customers, each of which have their own
damage functions. As illustrated in Chapters 3 and 4, the
primary adequacy indicators at HLII and HLIII are the
frequency and duration of load point interruptions. These
indices can be readily combined with a CCDF to give an
assessment of the expected customer losses due to interruption. The basic indices for adequacy assessment at
HLI are the loss of load expectation (LOLE) and the expected energy not supplied (EENS). These indices are illustrated in Chapter 2 using the RTS.

In order to assess reliability worth at HLI the CCDF
must be transformed into a cost factor which relates to
an acceptable adequacy index. This is known [15] as the

Interrupted Energy Assessment ·Rate (IEAR) and is discussed in the next sections.

Estimating The IEAR

Concepts. By far the broadest application of a customer damage function is its use in relating the composite customer losses to the socioeconomic worth of electric service reliability for an entire utility service
area. Reliability worth can be evaluated in terms of expected customer interruption cost. This cost estimate can
be obtained by multiplying the expected energy not supplied to customers due to power interruptions by a suitable factor. This factor, designated [15] as the Interrupted Energy Assessment Rate (IEAR), is expressed in
$/kWh. The expected energy not supplied is a basic gen-
erating system adequacy assessment index which can be
calculated in a number of ways. The severity and duration of discrete load loss events at HLI can be readily
assessed using the frequency and duration approach [14].
This method in conjunction with the appropriate customer
damage function can be used to estimate the IEAR.

Mathematical Formulation. The basic models required
in this approach are as follows:
(a) generation model - units are characterized by their
    capacity, forced outage rates, failure rates and re-
    pair rates,
(b) load model - the exact-state type load model [14] is
    used. This model represents the actual system load
    cycle by approximating it by a sequence of discrete
    load levels. The daily peak loads were assumed to
    exist for 12 hours giving an exposure factor of 0.5,
(c) cost model - this is represented either by the sector
    costs of interruption with their distribution of
    energy and peak demand for the service area or by the
    composite customer damage function.

The exact-state load model can be combined with an
exact-state capacity model to yield the frequency and duration associated with each load loss event [14]. The
Expected Energy Not Supplied (EENS) for each load loss
event is given by:

$$\text{EENS} = m_i f_i d_i \quad (\text{kWh/day}) \tag{5.1}$$

where m_i = margin state in kW of load loss event i,
      f_i = frequency in occ/day of load loss event i,
and d_i = duration in hours of load loss event i.

The probability, p_i of a load loss event i is given
by the product of the frequency, f_i (occ/day), and duration, d_i (days) of this load loss event. The total expected energy not supplied within the considered period
for all the load loss events is as follows:

$$\text{Total EENS} = \sum_{i=1}^{N} m_i p_i \tag{5.2}$$

where N is the total number of load loss events.

In deriving the cost associated with this unserved
energy, attention should be drawn to the units of the
composite customer damage function [3]. A major assumption in the estimation of the cost component is that the
magnitude of each load loss event is assumed to be related to the annual peak demand of a particular customer.
The total expected cost for all the load loss events of
the system is given by the following equation:

$$\text{Total expected cost} = \sum_{i=1}^{N} m_i f_i c_i(d_i) \quad (\text{\$/day}) \tag{5.3}$$

where $c_i(d_i)$ is the interruption cost in \$/kW for a dura-
tion d_i in hours of a load loss event i.

The frequency and duration method is used to evaluate
the probability, frequency and duration of each load loss
state. For each interruption duration, the corresponding

cost is obtained from the costs of interruption data or
the composite customer damage function.

$$\text{Estimated IEAR} = \frac{\sum\limits_{i=1}^{N} m_i f_i c_i(d_i)}{\sum\limits_{i=1}^{N} m_i f_i d_i} \quad (\$/\text{kWh}) \qquad (5.4)$$

Application To The RTS

System Data. The generation data of the RTS is given
in Appendix 1. The daily peak loads as obtained from the
RTS were arranged in descending order and then grouped in
class intervals. The mean of each class was taken as the
load level and the class frequency as the number of oc-
currences of that load level. This load model is shown
in Table 5.26. The period of study is 364 days and the
peak demand is 2714 MW.

Table 5.26 - Exact-state load data

Peak load level (MW)	No. of occurrences (days)
2714.00	12.0
2458.00	82.0
2191.00	107.0
1928.50	116.0
1608.30	47.0
1485.40	364.0

The various assumed sector costs of interruption for
the RTS system are given in Table 5.27 and their distribution of energy and peak demand are given in Table 5.28.
The composite customer damage function is given in Table
5.29. The interruption duration was extended to 16 hours
to cover all expected interruption durations. This extension was assumed to be of the same slope as that be
tween 4 hours and 8 hours. Figure 5.12 shows the complete composite customer damage function.

Table 5.27 – Interruption cost data deemed applicable to
           example service area (1985 $/kW)

User sector	Interruption duration				
	1 min	20 min	1 hr	4 hr	8 hr
Residential	0.004	0.09	0.56	5.17	15.47
Large users	1.47	3.52	5.68	12.91	18.97
Small users	0.72	3.38	5.27	19.66	29.37
Gov't & inst.	0.04	0.36	1.45	6.35	25.23
Commercial	0.86	5.29	16.54	58.66	148.00
Office buildings	4.65	9.61	20.50	67.00	116.12
Farm	0.03	2.80	14.00	116.80	326.80

Table 5.28 – Distribution of energy and peak demand of
           example service area

User sector	Energy (%)	Peak demand (%)
Residential	31.0	34.0
Large users	31.0	30.0
Small users	19.0	14.0
Gov't & inst.	5.5	6.0
Commercial	9.0	10.0
Office buildings	2.0	2.0
Farm	2.5	4.0

Table 5.29 – Composite customer damage function

Interruption duration	Interruption cost (1985 $/kW)
1 min	0.73
20 min	2.42
1 hr	5.27
4 hr	19.22
8 hr	41.45

Results. Two cases were studied for the RTS. The
first considered the entire system and the second
considered the individual sectors.

Case (a). The composite customer damage function
shown in Table 5.29 was used as the cost model. The
frequency and duration method [14] is used to generate
the variables m, p, f and d in Equations 5.1 and 5.2 for
each load loss state. Using the duration d, the cost
associated with the load loss event is obtained from the

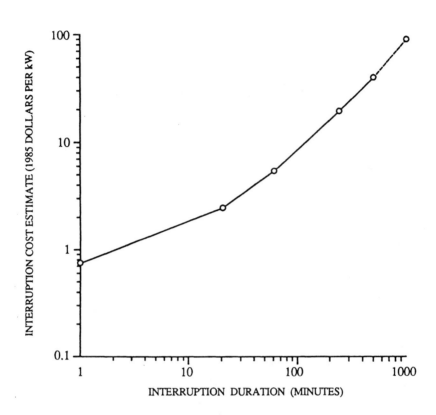

Figure 5.12 – Composite customer damage function for example service area

composite customer damage function. Equation 5.2 is used to evaluate the total expected unserved energy and the total expected interruption cost is obtained using Equation 5.3. Finally, Equation 5.4 is used to estimate the value of IEAR. In this example, the estimated IEAR is 5.02 $/kWh for an assumed exposure factor of 0.5.

The exact-state load model [14] is a function of the peak load and exposure factor. Studies were conducted to evaluate the impact of these variables on the estimated IEAR. First, the peak load was varied for a fixed exposure factor of 0.5. The results are shown in Figure 5.13. It can be seen that, a 125% change in peak load

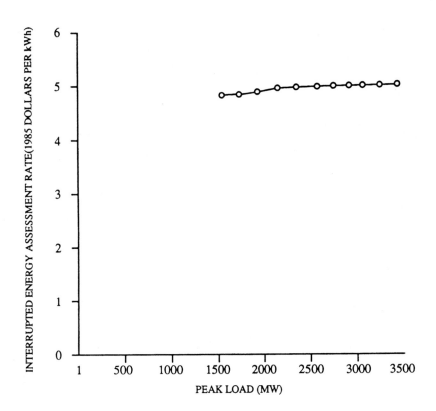

Figure 5.13 — Variation of IEAR with peak load

leads to only a 5% change in the estimated IEAR. It can
be concluded that the estimated IEAR for a particular
system (generation, load and composite customer damage
function) does not vary significantly with the peak load.

The effect of varying the exposure factor at a fixed
peak load is shown in Figure 5.14. A 350% change in exposure factor produces only about 6% change in the estimated IEAR. The estimated IEAR can therefore be considered to be relatively constant with variations in exposure factor at a fixed peak load.

The value of 5.02 $/kWh is therefore the best esti-
mate of the overall system IEAR for the RTS using the
composite customer damage function given in Table 5.29.

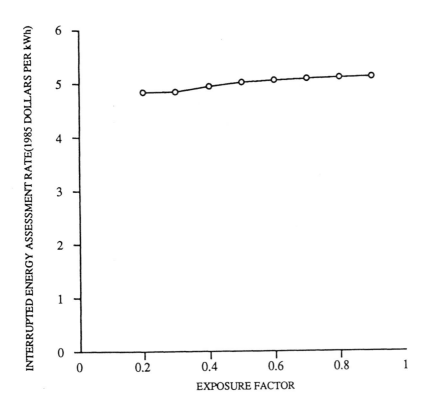

Figure 5.14 - Variation of IEAR with exposure factor

Case (b). The cost model utilized in the previous example was the composite customer damage function for the entire system. The analysis can also be conducted using each customer sector damage function. In order to illustrate this consider the residential sector. If the service area is made up of only the residential sector, then the residential interruption cost data given in Table 5.27 becomes the composite customer damage function of the service area. A residential sector IEAR can be calculated by applying the same procedure described in Case (a). A similar approach is employed to evaluate the IEAR of the other sectors within the service area. The

IEAR of each sector is weighted by its percentage of energy consumption to give that sector's expected IEAR.
The sum of the expected IEAR of all the sectors gives the
composite IEAR of the service area. The results are given
in Table 5.30. The estimated value of the IEAR for the
entire service area is 4.98 $/kWh.

Table 5.30 — Sector IEAR for an exposure factor of 0.5

User sector	Energy (%)	Sector IEAR ($/kWh)	Sector expected IEAR ($/kWh)
Residential	31.0	1.64	0.51
Large users	31.0	2.71	0.84
Small users	19.0	4.16	0.79
Gov't & inst.	5.5	2.39	0.13
Commercial	9.0	16.79	1.51
Office buildings	2.0	15.43	0.31
Farm	2.5	35.54	0.89
			4.98

Comments On Results. The IEAR values obtained from
the two cases under study are slightly different due to
interpolation of the respective customer damage functions. One advantage associated with the approach of
Case (b) is that the IEAR for the various sectors within
the service area of study are also available. These values can be used to assess the customer interruption costs
for any particular sector and can be used to analyze the
consequences associated with different load shedding
policies.

Application In Generating Capacity Planning

Conventional adequacy assessment at HLI is normally
done using a Loss of Load Expectation (LOLE) criterion.
As discussed in Chapter 2, there is an increasing awareness that an energy based index such as Expected Energy
Not Supplied (EENS) is a more responsive and perhaps a
more valid index. This section extends the utilization

of an energy based index by presenting a basic methodology for adding a monetary component to the conventional EENS. This factor, the IEAR, can be used in capacity adequacy assessment to produce an estimate of reliability worth. The comparison of reliability worth with reliability cost is an important component in the assessment and determination of an appropriate planning index.

Customer interruption costs have been used to analyze the impact of over/under capacity planning. The inherent uncertain nature of long term capacity planning creates a situation in which there is the potential for mismatch in supply and demand. An important component in the assessment of the implications associated with an inadequate supply is the customer costs associated with that condition. The IEAR as presented in this chapter can be tailored to any particular system and restricted to any particular class or classes of customers. It is response to the factors, such as the customers in the system, that should be present in any determination of what is an acceptable level of reliability. The actual magnitude of the IEAR will depend on the particular system under study. This chapter shows that a practical estimate can be obtained using the basic frequency and duration approach. The calculated results shown for the RTS suggest that a single IEAR value can be obtained and utilized in a wide range of system studies. When combined with a basic LOEE index, the IEAR provides a fundamental tool for assessing adequacy worth in HLI adequacy evaluation.

REFERENCES

1. Wacker, G., Wojczynski, E. and Billinton, R., "Cost/ Benefit Considerations In Providing an Adequate Electric Energy Supply," Third International Symposium on Large Engineering Systems, July 10-11, 1980. St. John's, Newfoundland, pp. 3-8.
2. Skof, L.V., "Ontario Hydro Surveys On Power System Reliability: Summary Of Customer Viewpoints," Ontario Hydro Report R&MR 80-12, EPRI Seminar, Oct. 11-13. 1983.

3. Billinton, R., Wacker, G. and Wojczynski, E., "Customer Damage Resulting From Electric Service Interruptions," Canadian Electrical Association, R&D
   Project 907 U 131 Report. 1982.
4. Wojczynski, E., Billinton, R. and Wacker, G., "Interruption Cost Methodology And Results - A Canadian
   Commercial And Small Industry Survey," IEEE Transactions PAS-103, 1983, pp. 437-444.
5. Wacker, G., Wojczynski, E. and Billinton, R., "Interruption Cost Methodology And Results - A Canadian
   Residential Survey," IEEE Transactions PAS-102, No
   10, October 1983, pp. 3385-3392.
6. Wacker, G., Subramaniam, R.K. and Billinton, R., "Using Cost Of Electric Service Interruption Surveys In
   The Determination Of A Composite Customer Damage
   Function," International Association of Science and
   Technology for Development Energy Symposia, June 4-6,
   1984. San Francisco, California, Paper No. 203-143.
7. IEEE Committee, "Report On Reliability Survey Of Industrial Plants, Part II, Cost Of Power Outages,
   Plant Restast Time, And Type Of Loads Lost Versus
   Time Of Power Outages," IEEE PAS, March/April 1974,
   pp. 236-241.
8. Ontario Hydro Survey on Power System Reliability
   Viewpoint of Customers in Retail Trade and Service OH
   Report No. R&U 79-7.
9. Ontario Hydro Survey on Power System Reliability:
   Viewpoint of Small Industrial Users (Under 5000 kw)
   OH Report No. R&V 78-3, April 1978.
10. Billinton, R., Wacker, G. and Subramaniam, R.K.,
    "Factors Affecting The Development Of A Residential
    Customer Damage Function," IEEE Transactions PWRS-2,
    No. 1, February 1982, pp. 204-209.
11. Billinton, R., Wacker, G. and Subramaniam, R.K.,
    "Factors Affecting The Development Of A Commercial
    Customer Damage Function," IEEE Transactions PWRS-1.
    No. 4, November 1986, pp. 28-33.
12. Subramaniam, R.K., Billinton, R. and Wacker, G.,
    "Factors Affecting The Development Of An Industrial
    Customer Damage Function," IEEE Transactions PAS-104,
    No. 11, November 1985, pp. 3209-3215.
13. Billinton, R., Hall, B.J. and Wacker, G., "A Program
    And Canadian Data Base For The Determination Of Indi-
    vidual And Composite Customer Damage Functions,"
    Proceedings 14th Inter-RAM Conference for the Elec-
    tric Power Industry, Toronto, June 1987.
14. Billinton, R. and Allan, R.N., "Reliability Evalua-
    tion Of Power Systems," Longman, London (England)/
    Plenum Press New York, 1984.
15. Billinton, R., Oteng-Adjei, J. and Ghajar, R., "Com-
    parison Of Two Alternate Methods To Establish An In-
    terrupted Energy Assessment Rate", IEEE Transactions,
    PWRS-2, No. 3, August 1987, pp. 751-757.

CHAPTER 6

CONCLUSIONS

The behavior of all engineering systems is essentially stochastic in nature, i.e. it varies randomly with time. Consequently, it is necessary to use models and analytical techniques that reflect this stochastic behavior in order to objectively evaluate future predictions. This requires the use of probabilistic assessment; to constrain the problem into a deterministic domain is unrealistic and prevents the effect of all system parameters to be quantitatively predicted. In fact, erroneous or misleading decisions can then ensue.

The need for probabilistic evaluation of power system behavior has been recognized since at least the 1930s, and it may be questioned why such methods have not been widely used in the past. The main reasons were lack of data, limitations of computational resources, lack of realistic techniques, aversion to the use of probabilistic techniques and a misunderstanding of the significance and meaning of probabilistic criteria and indices. None of these reasons need be valid today as most utilities have relevant reliability data bases, computing facilities are greatly enhanced, evaluation techniques are highly developed and most engineers have a working understanding of probabilistic techniques. Consequently, there is now no need to artificially constrain the inherent probabilistic or stochastic nature of a power system into a deterministic framework.

A wide range of probabilistic techniques have been developed. These include techniques for reliability

evaluation [1-6], probabilistic load flow [7] and probabilistic transient stability [8].   The fundamental and
common concept behind each of these developments is the
need   to   recognize   that   power   systems   behave
stochastically and all input and output state and event
parameters are probabilistic variables.   This book has
only been concerned with reliability evaluation and
reliability worth [9-11].

Present-day studies suggest that the "worst case"
system conditions which occur very infrequently should
not be utilized as design limits or criteria because of
economic restrictions. Probabilistic techniques have been
developed which recognize not only the severity of a
state or an event and its impact on system behavior and
operation, but also the likelihood or probability of its
occurrence.   Deterministic techniques can not respond to
the latter aspect and therefore can not account objectively for stochastic variables of the system.

An increasing number of utilities are now aware of
the benefits that can be derived from the use of quantitative reliability evaluation techniques.   Also, an increasing amount of material related to practical application of probability theory has been developed in recent
years.   It has been our intention to integrate both of
these aspects in this book so that the practical range of
techniques are documented together.   We have been very
selective in our choice of material and have neglected
that which we believe is still in the research domain or
which is not likely to have great practical significance.
Consequently, the material included in the previous chapters is already in use by utilities or can be used for
practical systems.

The RTS is now well established as a test system and
has been widely used in many diverse publications on a
worldwide scale.   We have therefore used it extensively
throughout this book.   This satisfies two distinct pur-

poses. Firstly, it serves as a consistent vehicle for demonstrating the different techniques and application areas. This shows what should and can be done in practical applications and the decisions that can be based on the results. Secondly, it increases the understanding of the RTS so that other studies have a basis of comparison. This we find to be of great importance since both of us have seen the RTS misused and misunderstood giving totally misleading and inappropriate indices.

Finally, increasing socioeconomic pressures to create safe and reliable power systems are being exerted on utilities by governments, environmental groups and society in general. We hope that the material presented in this book will play a significant role in finding acceptable solutions to such pressures and will encourage the increased use of reliability techniques in practical applications.

REFERENCES

1. Billinton, R. and Allan, R.N., "Reliability Evaluation Of Engineering Systems, Concepts And Techniques," Longman, London (England)/Plenum Publishers, New York, 1983.
2. Billinton, R. and Allan, R.N., "Reliability Evaluation Of Power Systems," Longman, London (England)/Plenum Publishers, New York, 1984.
3. Billinton, R., "Bibliography On The Application Of Probability Methods In Power System Reliability Evaluation," IEEE Transactions, PAS-91, pp. 649-660, 1972.
4. IEEE Committee Report, "Bibliography on the Application Of Probability Methods In Power System Reliability Evaluation, 1971-1977," IEEE Transactions, PAS-97, pp. 2235-2242, 1978.
5. Allan, R.N., Billinton, R. and Lee, S.H., "Bibliography On The Application Of Probability Methods In Power System Reliability Evaluation, 1977-1982," IEEE Transactions, PAS-103, pp. 275-282, 1984.
6. Allan, R.N., Billinton, R., Shahidehpour, S.M. and Singh C., "Bibliography On The Application Of Probability Methods In Power System Reliability Evaluation, 1982-1987," IEEE Winter Power Meeting, New York February 1988.
7. Allan, R.N., Leite da silva, A.M. and Burchett, R.C., "Evaluation Methods And Accuracy In Probabilistic

Load Flow Solutions," IEEE Trans. on Power Apparatus
and Systems, PAS-100, 1981, pp. 2539-2546.

8. Billinton, R. and Kuruganty, P.R.S., "A Probabilistic
   Index For Transient Stability," IEEE Transactions,
   PAS-99, pp. 195-207, 1980.

9. Billinton, R., Wacker, G. and Wojczynski, E., "Comprehensive Bibliography Of Electrical Service Interruption Costs," IEEE Transactions, PAS-102, pp.
   1831-1837, 1983.

10. Billinton, R., Wacker, G. and Wojczynski, E., "Cus-
    tomer Damage Resulting From Electric Service Inter-
    ruptions," Canadian Electrical Association, R&D
    Project 907 U 131 Report, 1982.

11. Wacker, G., Billinton, R. and Brewer, R.E., "Farm
    Losses Resulting From Electric Service Interrup-
    tions," Canadian Electrical Association, R&D Project
    309 U 403 Report, 1987.

APPENDIX 1

IEEE RELIABILITY TEST SYSTEM

INTRODUCTION

The IEEE Reliability Test System (RTS) was developed in order to provide a common test system which could be used for comparing the results obtained by different methods. The full details of the RTS can be found in the following reference: "IEEE Reliability Test System", Transactions on Power Apparatus and Systems, PAS-98, 1979, pp. 2047-2054.

This appendix presents some of the basic generation, load and network data which was defined for the RTS. The reader, however, is referred to the original reference for a greater discussion of this data and the bases for its creation.

LOAD MODEL

The annual peak load for the test system is 2850 MW.

Table A1.1 gives data on weekly peak loads in percent of the annual peak load. If week 1 is taken as January, Table A1.1 describes a winter peaking system. If week 1 is taken as a summer month, a summer peaking system can be described.

Table A1.2 gives a daily peak load cycle, in percent of the weekly peak. The same weekly peak load cycle is assumed to apply for all seasons. The data in Tables A1.1 and A1.2, together with the annual peak load define a daily peak load model of 52 x 7 = 364 days, with Monday as the first day of the year.

Table A1.1 - Weekly peak load in percent of annual peak

Week	Peak load	Week	Peak load
1	86.2	27	75.5
2	90.0	28	81.6
3	87.8	29	80.1
4	83.4	30	88.0
5	88.0	31	72.2
6	84.1	32	77.6
7	83.2	33	80.0
8	80.6	34	72.9
9	74.0	35	72.6
10	73.7	36	70.5
11	71.5	37	78.0
12	72.7	38	69.5
13	70.4	39	72.4
14	75.0	40	72.4
15	72.1	41	74.3
16	80.0	42	74.4
17	75.4	43	80.0
18	83.7	44	88.1
19	87.0	45	88.5
20	88.0	46	90.9
21	85.6	47	94.0
22	81.1	48	89.0
23	90.0	49	94.2
24	88.7	50	97.0
25	89.6	51	100.0
26	86.1	52	95.2

Table A1.2 -   Daily peak load in percent of weekly peak

Day	Peak load
Monday	93
Tuesday	100
Wednesday	98
Thursday	96
Friday	94
Saturday	77
Sunday	75

    Table A1.3 gives weekday and weekend hourly load mod-
els for each of three seasons.
    Combination of Tables A1.1, A1.2, and A1.3 with the
annual peak load defines an hourly load model of 364 x 24
= 8736 hours. The annual load factor for this model can
be calculated as 61.4%.

Table A1.3 - Hourly peak load in percent of daily peak

Hour	Winter weeks 1-8 & 44-52 Wkdy	Wknd	Summer weeks 18 - 30 Wkdy	Wknd	Spring/Fall weeks 9-17 & 31-43 Wkdy	Wknd
12-1 am	67	78	64	74	63	75
1-2	63	72	60	70	62	73
2-3	60	68	58	66	60	69
3-4	59	66	56	65	58	66
4-5	59	64	56	64	59	65
5-6	60	65	58	62	65	65
6-7	74	66	64	62	72	68
7-8	86	70	76	66	85	74
8-9	95	80	87	81	95	83
9-10	96	88	95	86	99	89
10-11	96	90	99	91	100	92
11-Noon	95	91	100	93	99	94
Noon-1 pm	95	90	99	93	93	91
1-2	95	88	100	92	92	90
2-3	93	87	100	91	90	90
3-4	94	87	97	91	88	86
4-5	99	91	96	92	90	85
5-6	100	100	96	94	92	88
6-7	100	99	93	95	96	92
7-8	96	97	92	95	98	100
8-9	91	94	92	100	96	97
9-10	83	92	93	93	90	95
10-11	73	87	87	88	80	90
11-12	63	81	72	80	70	85

Wkdy = Weekday,   Wknd = Weekend

GENERATING SYSTEM

Table A1.4 gives a list of the generating unit rat-

Table A1.4 - Generating unit reliability data

Unit size MW	Number of units	Forced outage rate	MTTF hr	MTTR hr	Scheduled maintenance wk/yr
12	5	0.02	2940	60	2
20	4	0.10	450	50	2
50	6	0.01	1980	20	2
76	4	0.02	1960	40	3
100	3	0.04	1200	50	3
155	4	0.04	960	40	4
197	3	0.05	950	50	4
350	1	0.08	1150	100	5
400	2	0.12	1100	150	6

ings and reliability data.

Table A1.5 gives operating cost data for the generating units. For power production, data is given in terms of heat rate at selected output levels. The following fuel costs are suggested for general use:

Table A1.5 – Generating unit operating cost data

Size MW	Type	Fuel	Output %	Heat rate Btu/kWh	O&M cost Fixed $/kW/yr	O&M cost Variable $/MWh
12	Fossil steam	#6 oil	20 50 80 100	15600 12900 11900 12000	10.0	0.90
20	Combus. turbine	#2 oil	80 100	15000 14500	0.30	5.00
50	Hydro					
76	Fossil steam	coal	20 50 80 100	15600 12900 11900 12000	10.0	0.90
100	Fossil steam	#6 oil	25 55 80 100	13000 10600 10100 10000	8.5	0.80
155	Fossil steam	coal	35 60 80 100	11200 10100 9800 9700	7.0	0.80
197	Fossil steam	#6 oil	35 60 80 100	10750 9850 9840 9600	5.0	0.70
350	Fossil steam	coal	40 65 80 100	10200 9600 9500 9500	4.5	0.70
400	Nuclear steam	LWR	25 50 80 100	12550 10825 10170 10000	5.0	0.30

#6 oil	$2.30/MBtu,
#2 oil	$3.00/MBtu,
coal	$1.20/MBtu,
nuclear	$0.60/MBtu.

TRANSMISSION SYSTEM

The transmission network consists of 24 bus locations connected by 38 lines and transformers, as shown in Figure A1.1. The transmission lines are at two voltages, 138 kV and 230 kV. The 230 kV system is the top part of Figure A1.1, with 230/138 kV tie stations at buses 11, 12, and 24.

The locations of the generating units are shown in Table A1.6.

Bus load data at time of system peak is shown in Table A1.7.

Transmission line forced outage data is given in Table A1.8.

Outages on substation components which are not switched as a part of a line are not included in the outage data in Table A1.8. For bus sections, the following data is provided:

	138 kV	230 kV
Faults per bus section-year	0.027	0.021
Percent of faults permanent	42	43
Outage duration for permanent faults, hr	19	13

For circuit breakers, the following statistics are provided:

Physical failures/breaker year	0.0066
Breaker operational failure per breaker year	0.0031
Outage duration, hr	72

A physical failure is a mandatory unscheduled removal from service for repair or replacement. An operational failure is a failure to clear a fault within the

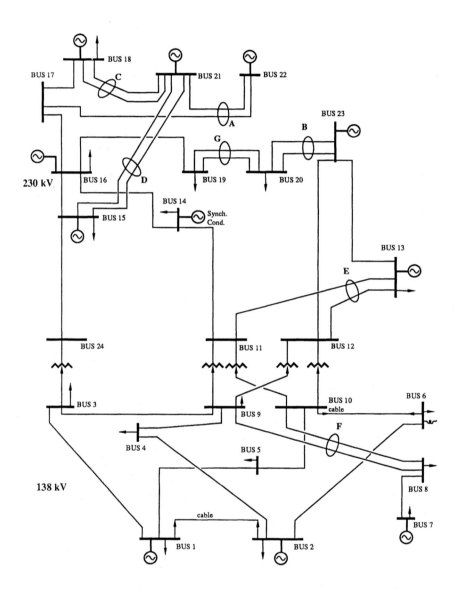

Figure A1.1 - IEEE reliability test system

Table A1.6 – Generating unit locations

Bus	Unit 1 MW	Unit 2 MW	Unit 3 MW	Unit 4 MW	Unit 5 MW	Unit 6 MW
1	20	20	76	76		
2	20	20	76	76		
7	100	100	100			
13	197	197	197			
15	12	12	12	12	12	155
16	155					
18	400					
21	400					
22	50	50	50	50	50	50
23	155	155	350			

Table A1.7 – Bus load data

	Load	
Bus	MW	MVAr
1	108	22
2	97	20
3	180	37
4	74	15
5	71	14
6	136	28
7	125	25
8	171	35
9	175	36
10	195	40
13	265	54
14	194	39
15	317	64
16	100	20
18	333	68
19	181	37
20	128	26
Total	2850	580

breaker's normal protection zone.

There are several lines which are assumed to be on a common right-of-way or common tower for at least a part of their length. These pairs of lines are indicated in Figure A1.1 by circles around the line pair, and an associated letter identification. Table A1.9 gives the actual length of common right-of-way or common tower.

Table A1.8 – Transmission line length and forced outage
            data

			Permanent		Transient
From bus	To bus	Length miles	Outage rate 1/yr	Outage duration hr	Outage rate 1/yr
1	2	3	.24	16	0.0
1	3	55	.51	10	2.9
1	5	22	.33	10	1.2
2	4	33	.39	10	1.7
2	6	50	.48	10	2.6
3	9	31	.38	10	1.6
3	24	0	.02	768	0.0
4	9	27	.36	10	1.4
5	10	23	.34	10	1.2
6	10	16	.33	35	0.0
7	8	16	.30	10	0.8
8	9	43	.44	10	2.3
8	10	43	.44	10	2.3
9	11	0	.02	768	0.0
9	12	0	.02	768	0.0
10	11	0	.02	768	0.0
10	12	0	.02	768	0.0
11	13	33	.40	11	0.8
11	14	29	.39	11	0.7
12	13	33	.40	11	0.8
12	23	67	.52	11	1.6
13	23	60	.49	11	1.5
14	16	27	.38	11	0.7
15	16	12	.33	11	0.3
15	21	34	.41	11	0.8
15	21	34	.41	11	0.8
15	24	36	.41	11	0.9
16	17	18	.35	11	0.4
16	19	16	.34	11	0.4
17	18	10	.32	11	0.2
17	22	73	.54	11	1.8
18	21	18	.35	11	0.4
18	21	18	.35	11	0.4
19	20	27.5	.38	11	0.7
19	20	27.5	.38	11	0.7
20	23	15	.34	11	0.4
20	23	15	.34	11	0.4
21	22	47	.45	11	1.2

Impedance and rating data for lines and transformers

is given in Table A1.10. The "B" value in the impedance
data is the total amount, not the value in one leg of the
equivalent circuit.

Table A1.9 – Circuits on common right–of–way or common
             structure

Right-of-way identification	From bus	To bus	Common ROW miles	Common structure miles
A	22	21	45.0	
	22	17	45.0	
B	23	20		15.0
	23	20		15.0
C	21	18		18.0
	21	18		18.0
D	15	21	34.0	
	15	21	34.0	
E	13	11		33.0
	13	12		33.0
F	8	10		43.0
	8	9		43.0
G	20	19		27.5
	20	19		27.5

Table A1.10 - Impedance and rating data

From bus	To bus	R	X	B	Normal	Short term	Long term	Equipment
			Impedance			Rating (MVA)		
		p.u./100 MVA base				Short	Long	
1	2	.0026	.0139	.4611	175	200	193	138 kV cable
1	3	.0546	.2112	.0572	"	220	208	138 kV cable
1	5	.0218	.0845	.0229	"	"	"	"
2	4	.0328	.1267	.0343	"	"	"	"
2	6	.0497	.1920	.0520	"	"	"	"
3	9	.0308	.1190	.0322	"	"	"	"
3	24	.0023	.0839		400	600	510	Transformer
4	9	.0268	.1037	.0281	175	220	208	138 kV line
5	10	.0228	.0883	.0239	"	"	"	"
6	10	.0139	.0605	2.459	"	200	193	138 kV cable
7	8	.0159	.0614	.0166	"	220	208	138 kV line
8	9	.0427	.1651	.0447	"	"	"	"
8	10	.0427	.1651	.0447	"	"	"	"
9	11	.0023	.0839		400	600	510	Transformer
9	12	.0023	.0839		"	"	"	"
10	11	.0023	.0839		"	"	"	"
10	12	.0023	.0839		"	"	"	"
11	13	.0061	.0476	.0999	500	625	600	230 kV line
11	14	.0054	.0418	.0879	"	"	"	"
12	13	.0061	.0476	.0999	"	"	"	"
12	23	.0124	.0966	.2030	"	"	"	"
13	23	.0111	.0865	.1818	"	"	"	"
14	16	.0050	.0389	.0818	"	"	"	"
15	16	.0022	.0173	.0364	"	"	"	"
15	21	.0063	.0490	.1030	"	"	"	"
15	21	.0063	.0490	.1030	"	"	"	"
15	24	.0067	.0519	.1091	"	"	"	"
16	17	.0033	.0259	.0545	"	"	"	"
16	19	.0030	.0231	.0485	"	"	"	"
17	18	.0018	.0144	.0303	"	"	"	"
17	22	.0135	.1053	.2212	"	"	"	"
18	21	.0033	.0259	.0545	"	"	"	"
18	21	.0033	.0259	.0545	"	"	"	"
19	20	.0051	.0396	.0833	"	"	"	"
19	20	.0051	.0396	.0833	"	"	"	"
20	23	.0028	.0216	.0455	"	"	"	"
20	23	.0028	.0216	.0455	"	"	"	"
21	22	.0087	.0678	.1424	"	"	"	"

APPENDIX 2

ADDITIONAL DATA FOR USE WITH THE RTS

INTRODUCTION

As described in Chapters 2 and 3 it is desirable that factors additional to those specified in the original RTS [1] be included in HLI and HLII evaluation studies. The following sections present additional RTS information prepared in order to encourage the use of common sets of data. These data were used in the studies [2,3] described in Chapters 2 and 3.

DERATED STATES

The 400 MW and 350 MW units of the RTS have been given [2] a 50% derated state. The number of hours in each state are shown in Table A2.1 and were chosen so that the EFOR [4] of the units are identical to the FOR specified in the original RTS [1].

Table A2.1 – Generating unit derated state data

Unit size MW	Derated capacity MW	SH(1) hr	DH(2) hr	FOH(3) hr	EFOR (4)
350	175	1150	60	70	0.08
400	200	1100	100	100	0.12

Notes: (1) SH = service hours,

        (2) DH = derated state hours,

        (3) FOH = forced outage hours,

        (4) EFOR = equivalent forced outage rate.

MAINTENANCE SCHEDULE

The suggested maintenance schedule is shown in Table
A2.2.   This schedule is Plan 1 of Reference 5.   It complies with the maintenance rate and duration of the
original RTS and was derived using a levelized risk criterion.

Table A2.2 – Maintenance schedule

Weeks	Units on maintenance			
1,2	none			
3-5	76			
6,7	155			
8	197	155		
9	197	155	20	12
10	400	197	20	12
11	400	197	155	
12,13	400	155	20	20
14	400	155		
15	400	197	76	
16,17	197	76	50	
18	197			
19	none			
20	100			
21,22	100	50		
23-25	none			
26	155	12		
27	155	100	50	12
28	155	100	50	
29	155	100		
30	76			
31,32	350	76	50	
33	350	20	12	
34	350	76	20	12
35	400	350	76	
36	400	155	76	
37	400	155		
38,39	400	155	50	12
40	400	197		
41,42	197	100	50	12
43	197	100		
44-52	none			

ADDITIONAL GENERATING UNITS

Additional gas turbines can be used [2] with the RTS
in order to reduce the LOLE of the system to a level fre-

quently considered acceptable. These additional units
are shown in Table A2.3. All other data may be assumed
to be identical to the existing gas turbines of the RTS.

Table A2.3 - Additional gas turbines

Unit size MW	Forced outage rate	MTTF hr	MTTR hr
25	0.12	550	75

LOAD FORECAST UNCERTAINTY

The load levels are assumed [2] to be forecasted with
an uncertainty represented by a normal distribution having a standard deviation of 5%. This is equivalent to a
load difference of 142.5 MW at the peak load of 2850 MW.
The discretised peak loads are shown in Table A2.4 for a
load model with 7 discrete intervals. The probability
values shown in this table can be evaluated using standard techniques [6].

Table A2.4 - Data for load forecast uncertainty

Std. deviations from mean	Load level MW	Probability
-3	2422.5	0.006
-2	2565.0	0.061
-1	2707.5	0.242
0	2850.0	0.382
+1	2992.5	0.242
+2	3135.0	0.061
+3	3277.5	0.006
		1.000

TERMINAL STATION EQUIPMENT

The extended single line diagram of the RTS is given
in Figure A2.1.

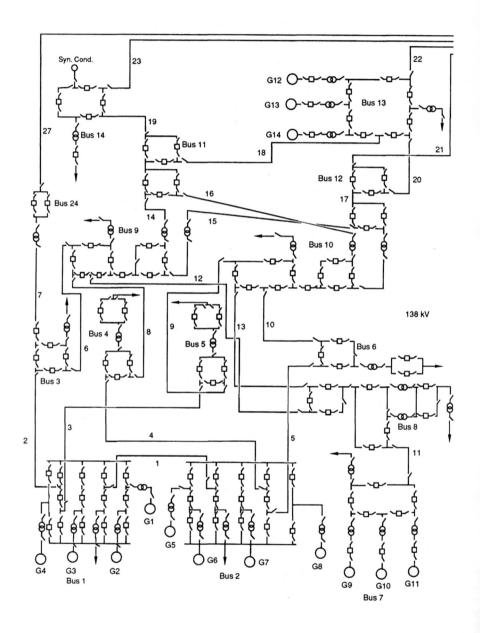

Figure A2.1 – Extended single line diagram of the RTS

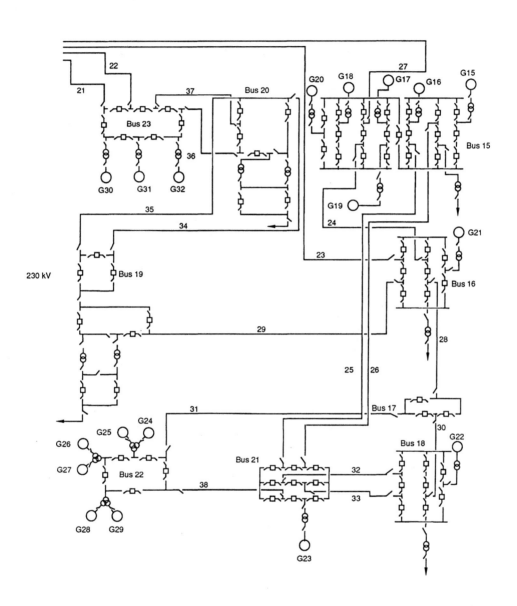

The additional data required to include the terminal stations [3] are as follows:

Active failure rate of a breaker   = 0.0066 f/yr,

Passive failure rate of a breaker = 0.0005 f/yr,

Maintenance rate of a breaker   = 0.2 outages/yr,

Maintenance time of a breaker   = 108 hr,

Switching time of a component   = 1.0 hr.

REFERENCES

1. IEEE Committee Report, "IEEE Reliability Test System," IEEE Trans. on Power Apparatus and Systems, PAS-98, 1979, pp. 2047-2054.
2. Allan, R.N., Billinton, R. and Abdel-Gawad, N.M., "The IEEE Reliability Test System - Extensions To And Evaluation Of The Generating System," IEEE Trans on Power Systems, PWSR-1, No. 4, 1986, pp. 1-7.
3. Billinton, R., Vohra, P.K. and Kumar, S., "Effect Of Station Originated Outages In A Composite System Adequacy Evaluation Of The IEEE Reliability Test System," IEEE Transactions PAS-104, No. 10, October 1985, pp. 2649-2656.
4. IEEE Std 762, "Definitions For Use In Reporting Electric Generating Unit Reliability, Availability And Productivity".
5. Billinton, R. and El-Sheikhi, F.A., "Preventive Maintenance Scheduling Of Generating Units In Interconnected Systems," International RAM Conference, 1983, pp. 364-370.
6. Billinton, R. and Allan, R.N., "Reliability Evaluation Of Engineering Systems, Concepts And Techniques," Longman, London, (England)/Plenum Press, New York, 1983.

APPENDIX 3

DEPENDENCY EFFECTS IN POWER SYSTEM RELIABILITY

INTRODUCTION

Although quantitative reliability evaluation is an
accepted aspect in the design and planning stage of many
systems, the most applicable evaluation method depends on
the type of system, its required function and the objective of the evaluation exercise. Most systems can be divided into one of two groups; mission-orientated systems
and continuously operated and repairable systems. Mission orientated systems include the aerospace industry
and safety applications and are, in general, concerned
with the probability of first failure. Repairable systems include the supply industries, electricity, gas,
etc., and continuous process plant and are, in general,
concerned with availability, i.e. the probability of
finding the system in an upstate at some future time.

The reliability evaluation of all systems, particularly power systems, is a very complex problem. The data
required to analyze this problem can be divided into two
basic parts as shown in Figure A3.1. In a simplistic
sense, these two requirements can be considered as deterministic data and stochastic data.

Deterministic data is required at both the system and
at the actual component level. The component data includes known parameters such as line impedances and sus-
ceptances, current-carrying capacities, generating unit
parameters and other similar factors normally utilized in
conventional load flow studies. This is not normally
difficult to determine as this data is used in a range of

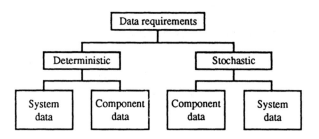

Figure A3.1 - Data requirements

range of studies. The system data, however, is more difficult to appreciate and to include and should take into
account the response of the system under certain outage
conditions. An example of this would occur if one of two
parallel lines suffered an outage; would the loading on
the remaining line be such that it would be removed from
service, would it carry the overload, or would some reme-
dial action be taken in the system in order to maintain
overall system integrity? The computer model must behave
in the same way as the actual system or the results are
not appropriate. This is an important aspect particularly
in composite system reliability evaluation as discussed
in Chapter 3 and is a problem that has not been properly
recognized up to this time.

Stochastic data can again be divided into two parts;
component and system data. The component requirements
pertain to the failure and repair parameters of the individual elements within the system. This data is generally available. There is also a need to consider and to
include system events which involve two or more components. This type of data is system specific and will
usually have to be inserted as a second and third level
of data input in an overall composite system reliability
analysis. System data includes relevant multiple failures resulting from dependent factors.

One significant assumption frequently made in the reliability evaluation of all systems, is that the behavior
of any one component is quite independent of the behavior
of any other component either directly or indirectly.   In
practice however it has been found that a system can fail
more frequently than the predicted value by a factor of
one, two or more orders of magnitude.   This effect is
known to be due to dependencies between failures of components, an effect which is neglected in the assumption
of component independence.   Two particular problems exist
before the effect of dependence can be included in reliability evaluation.   The first concerns recognition of
the modes and effects of dependence.   The second concerns
the   most   appropriate   reliability   model   and   evaluation
technique that accounts for these modes and effects.

Both of these aspects are discussed in this Appendix,
particularly in relation to continuously operated and repairable systems such as a power system.   The first is
dealt with in general terms only however because specific
details can not be described at a conceptual level.   The
second aspect is considered in terms of Markov modeling
techniques because the associated state space diagrams
give a clear and logical representation of the concepts
which are discussed.   Alternative evaluation techniques
can be used however provided the same overall concepts
are included.

CONCEPTS OF DEPENDENCIES

Classification

Independent outages are the easiest to deal with and
involve two or more elements.   They are referred to as
overlapping or simultaneous independent outages.   The
probability of such an outage is the product of the failure probabilities for each of the elements.   The basic
component model used in these applications is the simple
two-state representation in which the component is either

up or down. The rate of departure from a component upstate to its downstate is designated as the failure rate
λ. The restoration process from the downstate to the upstate is somewhat more complicated and is normally designated by the repair rate μ. The restoration of a forced
outage can take place in several distinct ways which can
result in a considerable difference in the probability of
finding the component in the downstate (usually designated as the unavailability). Some of the restoration
processes are:

(a)   high speed automatically re-closed,
(b)   slow speed automatically re-closed,
(c)   without repair,
(d)   with repair.

These processes involve different values of outage
times and therefore different repair rates. In addition
to forced outages, the component may also be removed from
service for a scheduled outage. The scheduled outage
rate, however, must not be added directly to the failure
rate as scheduled outages are not random events. For instance, the component is not normally removed from maintenance if the actual removal results in customer interruption. Most of the presently published techniques for
system reliability evaluation assume that the outages are
independent.

There are several modes of failures or effects that
can create dependence between the behavior of individual
components. Sometimes the underlying differences between
the various classifications have been confused and the
physical processes involved in the failure processes have
been inadvertently neglected. The main reason for this
is the need to include the effect of dependence at all
costs. It is however important to clearly understand the
processes involved and to choose the most appropriate
model in order to ensure that the evaluation responds to
and reflects the true system behavior. Failure to do so

can lead to misleading conclusions and misleading results. Dependencies are generally due either to some
common effect or cause, or to a cascading effect. The
main classifications include:

(a) common mode (or cause) failure,

(b) sharing a common environment such as weather,

(c) sharing common components including station originated effects,

(d) cascade failure,

(e) restricted repair and/or maintenance.

These are described in more detail in the following
sections.

Common Mode Failures

Considerable attention [1-12] has been given to the
effect of common mode failures in recent years; some being particularly concerned [1-3] with safety or mission-
orientated systems whilst others have been more concerned
[5-10] with repairable systems. Two papers by the authors
[10,12] considered this problem area in considerable
depth.

The essential aspect of a common mode failure is that
two or more components are outaged simultaneously due to
a common cause. One very satisfactory definition given
in Reference 4 (others also exist) is:

"a common mode failure is an event having a single
external cause with multiple failure effects which
are not consequences of each other.".

The most important features of such a definition are:

(a) the cause must be a single event,

(b) the single cause produces multiple effects. This
means that more than one system component is affected,

(c) the effects are not consequences of each other. This
means that all the components involved are affected
directly by the cause and not indirectly due to other

components having failed because of the cause.

System events that create conditions which cause other components to fail should be classed as consequential or cascade failures. These are discussed in a later section.

A detailed discussion of the significance of the above definition and typical practical examples was given in Reference 10. One primary requirement of a common mode failure which can be discerned from this definition is that the cause or initiating event is external to the system being analyzed. Consequently a cause should not be considered external, and therefore a common mode event, simply because the system boundary has been drawn around a restricted part of the system. In this case, the cause has been artificially made to appear as an external event whereas the system function or component causing it is a real internal part of the system but is not fully represented in the topology of the system being analyzed. Further consideration of this is given later.

Two particular examples that can be given to illustrate common mode events are:

(a) a single fire in a nuclear reactor plant causes the failure of both the normal cooling water system and the emergency cooling water system because the pumps for both systems are housed in the same pumping room,

(b) the crash of a light aircraft causes the failure of a two-circuit transmission line because both lines are on the same towers.

These two examples are useful because they illustrate two extremes in practical system operation.

In the first example, the system should be designed so that this type of common mode failure either can not occur or its chance of occurrence is minimized. The main function of a reliability assessment is to highlight this type of event and to establish its probability of occurrence. If the probability of the event is unacceptable, the system should be redesigned.

In the second example, the common mode failure event
itself may have to be accepted because environmental constraints force designers to use common towers and common
rights-of-way. The main function of the reliability assessment is to establish the significance of the event in
order to determine what other actions need to be taken to
minimize its effect.

These two examples are good illustrations of the
practical problem encountered with common mode failures
because they assist in identifying the two classes into
which most failures can be grouped. The first group, example (a), are those common mode failures which must be
identified and eliminated if at all possible. The second
group, example (b), are those common mode failures which
may have to be accepted but their effect must be minimized.

Other examples have been suggested as causes of common mode failures. These have included the same manufacturer, the same environment, the same repair team, etc.
These additional examples are frequently not common mode
failure events.

As an example, consider the case of a common manufacturer. The reason for its suggested inclusion is that a
product from a particular manufacturer may contain the
same essential weakness. This neglects the fact that the
failure process of each similar component is still independent albeit with a failure rate that may be significantly greater than that of a similar component from
another manufacturer. The cause of an overlapping failure event is therefore related to independent component
outages due to perhaps excessively high individual failure rates. There is no denying the importance of this
problem and the need to recognize it. It may therefore
be necessary to identify such events in a properly structured reliability evaluation and to employ diversity if

possible. On the other hand it is also necessary to ensure the correct interpretation of the failure process
and to use an appropriate reliability evaluation method
which correctly simulates this failure process.

The same principle applies to other causes relating
to system misuse by the operator or an inadequate repair
process by the repair team. In both cases, the effect is
to enhance the failure rate of the components. The failure process itself still involves independent overlapping
outages.

Common Environment

As discussed in the previous section, common changes
of environment have sometimes been classified as a common
mode event particularly those environments which have a
significant impact on the failure process. This is an
over-simplistic viewpoint and again neglects positive
consideration of the actual failure process itself.

It is found in practice that the failure rate of most
components are a function of the environment to which
they are exposed. In some adverse environments, the
failure rate of a component can be many times greater
than that found in the most favorable conditions. During
the adverse environmental conditions, the failure rates
increase sharply and the probability of overlapping failures is very much greater than that which occurs in favorable conditions. This creates a situation known as
[13] "bunching" due to the fact that component failures
are not randomly distributed throughout the year but are
more probable in constrained short periods. This bunching effect has been inadvertently construed as common
mode conditions whereas the bunching effect does not imply any dependence between the failures of components.
It simply implies that the component failure rates are
dependent on the common environment. There is therefore
no suggestion that the process is a common mode failure,

only that the independent failure rates are enhanced because of the common environment.

Two particular examples of this type of environmental
failure process are:

(a) weather conditions affecting the failure process of a
    double-circuit transmission line.   During adverse
    weather, e.g. gales, lightning storms, etc., the
    failure rates of each circuit are enhanced greatly
    thus increasing the probability of an independent
    overlapping outage,

(b) the normal cooling water system and emergency cooling
    water system of a nuclear reactor plant use pumps ex-
    posed to the same temperature or stress conditions.
    During adverse temperature or stress conditions, the
    failure rates of each pump are enhanced thus increas-
    ing the probability of an independent overlapping
    outage.

Common Components

Before the reliability of any system can be evaluated, it is first necessary to define the system and to
draw a system boundary around it. Very few systems can
be completely defined and totally encompassed by a system
boundary. Consequently, either some parts of the system
are not represented in full or some related part of the
system is left outside of the system boundary. This
simplification is sometimes necessary in order to make
the problem a tractable one. An example of the second
simplification is that, during the reliability assessment
of a process plant, the power supply to the plant is only
represented as a single injection and the electricity
supply system is not otherwise represented. It is of
course not practical to represent the whole of the power
supply although failures within it can have major effects
on the operation of the process plant.

Since failures within the parts of the system not

fully represented may cause two or more represented components to fail, it is tempting to classify such failures
as common mode failures since they cause multiple failure
effects. This concept has been used in system reliability evaluation. It is not correct to do so however because the initiating failure event is not external to the
system being analyzed and only appears to be external due
to the construction of the system boundaries.

One particularly important area of activity associated with sharing common components, is that of station
originated outages. These can have significant impact on
the behavior of composite generation/transmission systems
and are discussed separately in the next section.

Station Originated Outages

The overall reliability assessment of a composite
power system, if every possible system state is analyzed
and all types of system component outages are included,
involves an exhaustive and formidable analytical and computational effort. Therefore, in such studies, several
simplifications have been implemented to make such analysis less demanding. Such simplifications require a thorough knowledge of the system behavior and must be care-
fully assessed in order to avoid making assumptions which
would produce an unrealistic evaluation of the system
reliability.

One of the main simplifications is that terminal stations are modeled only as single busbars without considering the internal configuration of the station. Therefore, the internal failures of the stations, which could
have a serious effect on the system performance, are neglected.

This modeling procedure is illustrated using the
switching station shown in Figure A3.2. This figure
shows the single-line diagram of a ring-type station including the breakers, transformers and busbars which was

used [14] for modeling busbar 23 of the RTS (see Appendix
2). It also shows the generators and lines connected to
the station. The simplified method of representing this
station in a composite system reliability evaluation
technique would be as shown in Figure A3.3; i.e. a single
busbar to which the generators and lines are connected.

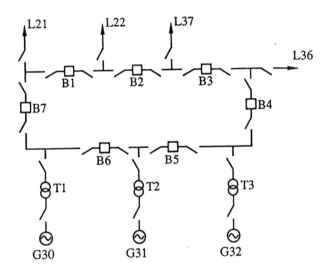

Figure A3.2 — Ring-bus station

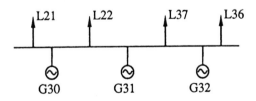

Figure A3.3 — Single bus representation

If considered, the outage effects of system components due to terminal station events have been included
by adding a factor to the failure and repair rates of
generators and/or lines affected by the failure of the
station component. This approach may be correct for out-

ages involving only one generator or line. However, for
states with more than one system component out, this procedure assumes independence of the outages of each individual component. This is frequently unrealistic.

The unsuitability of this approach can be demonstrat-
ed considering the station shown in Figure A3.2. Assume
that breaker B1 suffers a short-circuit which causes the
operation of breakers B2 and B7. This leads to the dis-
connection of two lines (L21 and L22); a second-order
contingency. The probability of this contingency is not
the product of the outage probabilities of L21 and L22
but the probability that a short-circuit occurs on B1.

Another simplification usually adopted is that a reliability study of composite systems concludes with the
analysis of outages up to a certain contingency level.
The assumption is that the probability of a state with a
high number of system components on outage is so small
that it can be neglected. This is correct if the outages
of the system components are considered as independent
events. However, this state can exist, not only due to
the overlapping occurrence of independent failures, but
due to a single failure of a station component; its probability of occurrence is therefore much greater.

These events are now known as station originated outages [14-17] where a station-originated event is defined
as a forced outage of any number of system generators and
lines, caused by a failure inside a switching or terminal
station. Therefore, these outages are dependent on the
occurrence of one or more failure events and their reliability indices are the values associated with the indices of the station components which fail.

Cascade Failures

There are many examples in practice where the failure
of one component enhances the stresses imposed on other
components. Under these circumstances, the failure rates

of the other components are increased and the probability
of failure made more likely. This process is generally
known as cascade or consequential failures and is due to
the fact that failure rates are stress dependent. This
concept is different to the case of environmental failures discussed previously since, in the previous section,
the failure rates of components were directly related to
the environmental conditions and independent of the state
of other components. In the present situation of cascade
failures, the component failure rates are dependent on
the state of other components.

A cascade failure may involve a sequence of events,
i.e. a chain of events may occur, each of which involves
the failure of one or more components; subsequent component failures being made more likely due to the increased
stress imposed upon them due to previous failures in the
sequence. At some point in the sequence, two or more
components may fail simultaneously due to the failure of
a previous component. This again appears to be a common
mode failure. This is not a correct interpretation however because it is due to internal cascade events and
therefore is akin to the sharing of components as described in a previous section.

A suitable definition of a cascade type of dependent
failure is: "A failure is a dependent failure if the occurrence of another failure event affects its probability
of occurrence. A dependent failure can therefore only
occur in association with an independent failure and must
be related to this independent failure.".

Because a dependent failure must be related to an independent event, the probability of occurrence is conditional. This type of event is therefore best considered
using standard conditional probability theory [18].

A particular example of a cascade failure is the col-
lapse of an electricity supply system due to the loss of
one circuit during high load conditions which imposes ex-

cessive load on the remaining circuits.

General Discussion

The previous discussion has been directed at identifying types of failure modes, causes and effects that can
occur in real systems and which can significantly increase the failure probability of the overall system.
The examples given are merely illustrative in order to
place the types of failure in perspective. Many more examples could be given but the reader should be able to
examine his own system in the light of the examples
quoted and identify into which category possible failure
events should be put.

The most important concept that should be concluded
from the previous discussion is that there are several
separate types of dependence categories and that it is
not realistic to define only two groups of failure processes; independent failure events and common mode failure events. The reason for these multiple categories is
due to the fact that the underlying failure process is
different in each case. The failure process involved in
any given system must be fully understood and appreciated
by the engineer involved in the reliability evaluation
and the most appropriate reliability model used. These
models are described in the following sections from which
it can be seen that the model for each category contains
specific and significant differences. To absorb all dependent failure events into one category, such as common
mode failures, is a manipulative exercise only and fails
to recognize the need that system models should reflect
the real behavior of the system.

RELIABILITY MODELING

Concepts Of Modeling

The models described in this section are illustrative

only. They are based on state space diagrams which can
be used as the input to a Markov analysis technique [18].
This method of illustrating the reliability models is
very useful since it clearly identifies, in a pictorial
form, the difference between the various failure categories described previously. It should be noted however
that the models are not rigid and should be modified as
required to suit the specific characteristics of any system being analyzed.

In addition, the models are generally developed in
terms of two components only. The concepts can however
be developed to any level of complexity although a two
component representation may be all that is necessary if
a network reduction technique is being used or the models
are employed in conjunction with the minimal cut set approach [18].

Common Mode Failures

Models that represent common mode failures in repairable systems have been described [7] in several papers
and have been previously discussed in depth by the authors [10]. The two basic models [7] for a two component
system or second order minimal cut set are shown in Figure A3.4 in which λ_c represents the common mode failure
rate. The difference between these two models is that
one has a single down state (Figure A3.4a) and the other
has two separate down states; one associated with independent failures, the other associated with common mode
failures (Figure A3.4b). These models can be compared
with that shown in Figure A3.5 for the same system but
when independent failures only can occur.

These models can be adapted to suit the requirements
of any particular system. For instance, μ_c in Figure
A3.4a may be zero if all repairs are done independently
or all repair transitions can be neglected if the model
represents a mission orientated system that is not re-

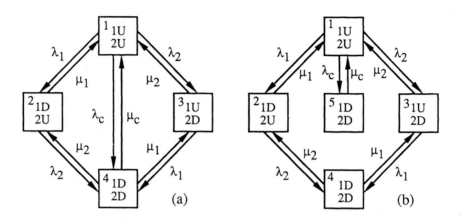

Figure A3.4 – Common mode failures

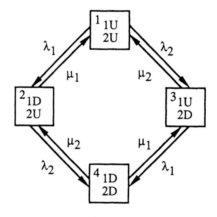

Figure A3.5 – Independent failure events

pairable. The models shown in Figure A3.4 therefore
represent the concept of common mode failures and can be
adapted at will.

Common Environment

As described previously, the failure process of each
component due to the environment is independent of that
of all other components. Consequently the independent

failure model shown in Figure A3.5 is the basis of the
model for common environmental considerations. Since the
failure rate of each component is different in each environment, a model similar to that of Figure A3.5 is required for each identified environmental condition, and
each of these models are interconnected by the appropriate environmental transition rates. This is shown [19]
in Figure A3.6 for a two component system which can exist
in one of two environmental states, normal and adverse.
In this model, λ_n represents the transition rate between
normal and adverse environment, λ_a represents the reverse
transition, λ_i represents the failure rate of component i
per year of normal environment and λ_i' represents the
failure rate of component i per year of adverse environment.

This model [19] which is simply a transformation of
the knowledge of system behavior into a reliability diagram, is clearly very different from that representing
common mode failures as illustrated in Figure A3.4. The
two aspects, common mode failures and common environmental effects, can not therefore be construed to be similar.

The reliability model shown in Figure A3.6 can also
be modified to suit particular system characteristics.
For instance, the repair process in both environments may
be similar or dissimilar, repair may not be possible during adverse environment, more than two environmental
planes can be included, common mode failures can be added
by including transitions between states 1 and 4 and
states 5 and 8.

A full discussion of these effects, models, evaluation techniques and relevant equations is given in References 13 and 18.

Common Components

In order to illustrate the modeling of commonly

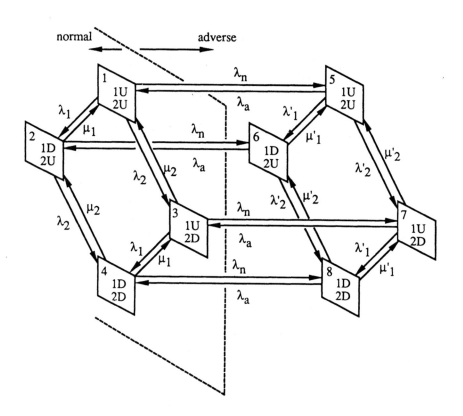

Figure A3.6 – Common environment

shared components, consider a two component system or
second order minimal cut set. The independent failure
modes for this order of event are represented by the
model shown in Figure A3.5. Consider now that a third
component, not included in the system, can outage both of
these components if it should fail. It could be presumed
that the model of Figure A3.5 can be modified to account
for this third component effect by adding transitions directly between states 1 and 4. This would produce a model
similar in structure to that of Figure 3.4b in which λ_c
is replaced by the failure rate of the third component λ_3
and μ_c is replaced by the repair rate of the third component μ_3. This however need not be correct and should

only be used if the correct model can be shown to reduce
to Figure A3.4b by justified simplifications of the system behavior.   The model of Figure A3.4a will never be
applicable because of the transitions that emanate from
the single "both components down" state.

A more realistic and comprehensive model for the
present example is shown in Figure A3.7.   In this model,
the state of component 3 is positively identified and all
possible states are included.   The up state for a parallel system or second order minimal cut set are 1-3 and
the down states are 4-8.   This model can be both extended
and simplified depending on the features which must be
included.

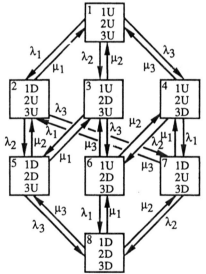

Figure A3.7 — Sharing a common component

The model can be extended by including real common
mode failures involving components 1 and 2.   This would
involve additional transitions between states 1 and 5 and
states 1 and 8. The model can also be extended to include
common environmental conditions by including a further
plane similar in structure to Figure A3.7 and identical

component states joined together by the environmental
transition rates.

Station Originated Outages

This section describes some possible models for
studying the effects of station originated outages.
These models can be used to account for outages affecting
two elements and can be extended to more than two elements [15].

The simplest possible model can be obtained by combining the station originated outages which results in
both lines out with the common-cause outages. This model,
which is shown in Figure A3.8a, is similar to that of

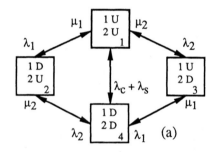

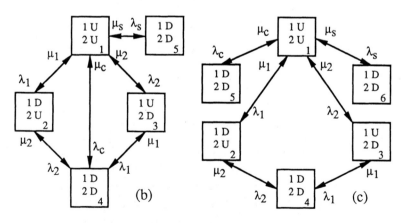

Figure A3.8 – Station originated outages: (a) Model 1,
(b) Model 2, (c) Model 3

Figure A3.4a except for the fact that the transition rate
from the both lines upstate to the both lines downstate
has been increased from λ_c to $\lambda_c + \lambda_s$ (where λ_s is the
contribution of station originated outages). This model
therefore assumes the same repair process for independent, common-cause, and station originated outages. A
serious objection to this model is the fact that the repair duration for a station originated outage will be
very short compared to the repair duration for common-
cause outages, and independent outages.

Another possible model is shown in Figure A3.8b. This
model is a simple modification of Model 1. In this model,
a separate state is created to account for the station
originated outages. A more practical model is shown in
Figure A3.8c. In this model, the all lines downstates
due to independent outages, common-cause outages and
station-originated outages are shown as three separate
states. Other models can be created to suit the data and
needs of a particular situation.

In Models 1 and 2, λ_c will be equal to zero if the
station originated outage involves a transmission line
and a generator, or two generators which can be considered to be independent. Under such circumstances, state
5 of Model 3 does not exist.

Cascade Failures

The concept of cascade failures implies that the
failure rate of a component is greatly enhanced following
the failure of another component due to the greater
stress imposed on the component. The consequence of this
is that each component must be defined by more than one
value of failure rate; its failure rate assuming independent failure processes and the failure rates that are
conditional on the previous failure of other components.

Again consider a two component system or second order
minimal cut set. Let λ_1 and λ_2 be the independent fail-

ure rates, λ_{1c} be the failure rate of component 1 conditional on the fact that component 2 has previously failed
and λ_{2c} be the failure rate of component 2 conditional on
the fact that component 1 has previously failed. In common with the definition of a transition rate, these values must be expressed in failures per unit time of being
in that state. Consequently, if the second failure is
very probable, the conditional failure rates of the two
components will be very large. Under these circumstances
of cascade failures, the model shown in Figure A3.5 can
be modified to that shown in Figure A3.9. This model can
be extended to include common mode failures and environmental considerations.

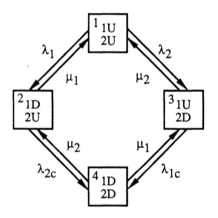

Figure A3.9 - Cascade failures

Other Dependent Effects

The considerations given in the previous sections are
specific and relate to particular system conditions and
outcomes. They do however indicate the considerations
that must be made in order to establish the most appropriate reliability model. A logical appreciation of the
previous considerations should therefore be beneficial in
structuring reliability models for other dependencies.

For instance, the previous models generally assumed that
repair was always possible and that no restrictions were
imposed on the amount of manpower available. Consequently, a state containing more than one failed component
could be departed by completing repair on any of the
components which were in the failed state, i.e. repair of
all failed components was being conducted simultaneously.

In practice this may not be physically possible because of limited manpower resources. In this case, only
a limited number of repairs are possible at the same time
and other components must wait their turn. The previous
models can be adapted to include these effects quite
readily. Consider for instance, the two component system
described by the model of Figure A3.5. Assume now that
only one component can be repaired at any one time and
that the component which fails first is repaired first.
The model shown in Figure A3.5 must be modified by dividing the "both components down" state into two substates
and inserting the appropriate transitions between states.
This adaptation is shown in Figure A3.10.

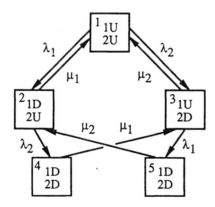

Figure A3.10 – Restricted repair

Similar adaptations can be made to all of the previ-

ous models in order to account both for this repair dependence and any other dependent characteristic known to
exist in the system.

EVALUATION TECHNIQUES

After a reliability model has been constructed, a
suitable evaluation technique must be used to assess the
reliability of the system. There are a number of techniques that can be used. It is not the purpose of this
Appendix to describe such techniques in any detail because these are already well documented [13,18]. One way
is to construct a stochastic transitional probability
matrix from the state space diagram and to solve this using Markov techniques [18].

A second method, which proves very convenient in conjunction with minimal cut set analysis, is to deduce a
set of approximate equations from the reliability model
into which the appropriate transition rates can be inserted. Such equations [6,13,18,20] exist for common mode
failures and for environmental considerations [9,13,21].
In the latter case these were derived for the weather
effects of transmission lines but they can be used equally well for any other environmental condition.

CONCLUSIONS

This appendix has described some of the fundamental
dependence conditions that can arise in a system and how
these effects can be included in a reliability model of
the system. Several important concepts have been contained in the considerations discussed; these being:
(a) a thorough understanding of the system, its opera-
    tional characteristics, its failure modes and the
    likely cause of dependence is required before any re-
    liability model can be constructed,
(b) grouping of dependence effects into a restricted num-
    ber of categories, for example, common mode failures,

can create inappropriate reliability models which
subsequently do not reflect or respond to the real
behavior of the system,

(c) it is generally possible to construct a reliability
model for any system dependency and subsequently to
analyze it using existing evaluation methods.   The
greatest difficulty is in recognizing whether such a
dependency exists.   No technique can ever be developed that will remove from the analyst the need to
fully appreciate the modes of failures of his system,

(d) most dependencies are system-specific, i.e. they exist in specific systems to a greater or lesser extent
depending on the characteristics and operating conditions of the system.   Consequently they cannot be
generalized and it is not possible to create a
"black-box" that can simulate the dependence characteristics in a generalized way.   Such dependencies
must therefore be constructed specifically for each
system being analyzed.

REFERENCES

1. Fussell, J.B., Burdick, G.R. (eds), "Nuclear Systems
   Reliability Engineering And Risk Assessment," SIAM,
   1977.
   (a)   Epler, E.P., "Diversity And Periodic Testing In
   Defence Against Common Mode Failure," pp. 269-288.
   (b)   Wagner, D.P., Cate, C.L. and Fussell, J.B.,
   "Common Cause Failure Analysis Methodology For Complex Systems," pp. 289-313.
   (c)   Vesely, W.E., "Estimating Common Cause Failure
   Probabilities In Reliability And Risk Analyses," pp.
   314-341.
2. Epler, E.P., "Common Mode Failure Considerations In
   The Design Of Systems For Protection And Control,"
   Nuclear Safety, 10, 1969, pp. 38-45.
3. Edwards, G.T. and Watson, I.A., "A Study Of Common
   Mode Failures," National Center of Systems Reliability, Report SRD R146, 1979.
4. Gangloff, W.C., "Common Mode Failure Analysis," IEEE
   Trans. on Power Apparatus and Systems, PAS-94, 1970,
   pp. 27-30.
5. Billinton, R., Medicherla, T.K.P. and Sachdev, M.S.,
   "Common Cause Outages In Multiple Circuit Power
   Lines," IEEE Trans. on Reliability, R-27, 1978, pp.

        128-131.
   6. Allan, R.N., Dialynas, E.N. and Homer, I.R., "Model-
      ing Common Mode Failures In The Reliability Evalua-
      tion Of Power System Networks," IEEE PES Winter Power
      Meeting, New York, 1979, paper A79 040-7.
   7. Billinton, R., "Transmission System Reliability Mod-
      els," EPRI Publication, WS-77-60, pp. 2.10-2.16.
   8. Task Force of the IEEE APM Subcommittee, "Common Mode
      Forced Outages Of Overhead Transmission Lines," IEEE
      Trans. on Power Apparatus and Systems, PAS-95, 1976,
      pp. 859-864.
   9. Billinton, R. and Kumar, S., "Transmission Line Reli-
      ability Models Including Common Mode And Adverse
      Weather Effects," IEEE PES Winter Power Meeting, New
      York, 1980, paper A80 080-2.
  10. Allan, R.N. and Billinton, R., "Effect Of Common Mode
      Failures On The Availability Of Systems," 6th Advanc-
      es in Reliability Technology Symposium, NCSR Report
      No. R23, Vol. 2, July 1980, pp. 171-190.
  11. Watson, I.A., "Review Of Common Cause Failures," NCSR
      Report No. R27, 1981.
  12. Allan, R.N. and Billinton, R., "Effect Of Common
      Mode, Common Environment And Other Common Factors In
      The Reliability Evaluation Of Repairable Systems,"
      7th Advances in Reliability Technology Symposium,
      Bradford, 1981, paper 4B/2.
  13. Billinton, R. and Allan, R.N., "Reliability Evalua-
      tion Of Power Systems," Longman, (London)/Plenum (New
      York), 1984.
  14. Billinton, R., Vohra, P.K. and Kumar, S., "Effect of
      Station Originated Outages In A Composite System Ad-
      equacy Evaluation Of The IEEE Reliability Test Sys-
      tem," IEEE Transactions on PAS, PAS-104, 1985, pp.
      2649-56.
  15. Billinton, R. and Medicherla, T.K.P., "Station Origi-
      nated Multiple Outages In The Reliability Analysis Of
      A Composite Generation And Transmission System," IEEE
      Transactions on PAS, PAS-100, 1981, pp. 3870-78.
  16. Allan, R.N. and Adraktas, A.N., "Terminal Effects And
      Protection System Failures In Composite System Reli-
      ability Evaluation," IEEE Transactions on PAS, PAS-
      101, 1982, pp. 4557-62.
  17, Allan, R.N. and Ochoa, J.R., "Modeling And Assessment
      Of Station Originated Outages For Composite Systems
      Reliability Evaluation," IEEE Winter Power Meeting,
      New Orleans, 1987, paper 87 WPM 016-9.
  18. Billinton, R. and Allan, R.N., "Reliability Evalua-
      tion Of Engineering Systems; Concepts And Tech-
      niques," Longman, (London)/Plenum (New York), 1983.
  19. Billinton, R. and Bollinger, K.E., "Transmission Sys-
      tem Reliability Evaluation Using Markov Processes,"
      IEEE Trans. on Power Apparatus and Systems, PAS-87,
      1968, pp. 538-547.
  20. Allan, R.N., Avouris, N.M., Kozlowski, A. and
      Williams, G.T., "Common Mode Failure Analysis In The

Reliability Evaluation Of Electrical Auxiliary Systems," IEE Conference on Reliability of Power Supply
Systems, London, 1983, IEE Conf. Publ. 225, pp.
132-136.
21. Billinton, R. and Grover, M.S., "Reliability Assessment Of Transmission And Distribution Schemes," IEEE
Trans. on Power Apparatus and Systems, PAS-94, 1975,
pp. 724-733.

APPENDIX 4

EVALUATION OF STATISTICAL DISTRIBUTIONS

INTRODUCTION

This analytical approach [1] utilizes the first four moments of a reliability index to evaluate its percentiles. The analysis requires three major steps. In the first step, the first four raw moments of component failure and repair times and the system restoration times are determined. In the second step, the average value and the second, third and fourth central moments of the reliability indices are evaluated using the moments obtained in the first step and the information regarding the system configuration. The last step utilizes the Pearson method to evaluate the approximate percentiles of the reliability indices.

NOTATION

The following notation is used in the analysis [1].

General

$E[X_i^r]$ rth raw moment of X_i. The $E[X_i^r]$ for $r = 1$ e.g. $E[X_i]$ denotes the mean value of X_i,

$\mu_r[X_i]$ rth central moment of X_i.

Summation Indices (Subscripts)

x   includes all those components which if any one fails results in an interruption of at least one load point in the system,

y   includes all load points in the system,

z   includes all possible modes of restoration be-
                   sides components repairs, in the system,

$xy\varepsilon x$ includes those components which if any one
                   fails results in an interruption at load point
                   y,

$yx\varepsilon y$ includes those load points which will experi-
                   ence an interruption if the component x fails,

$yxo\varepsilon yx$ includes those load points which are restored
                   by repairing the failed component x,

$yxz\varepsilon yx$ includes those load points which are restored
                   by mode of restoration z, after experiencing an
                   outage due to the failure of component x.

Deterministic Quantities

c_y number of customers connected at load point y,

l_y total average load (kW) connected at load point
                   y,

c_t total number of customers connected to the sys-
                   tem,

l_t total average load on the system.

Random Variables

P_x probability that a given component failure
                   event in the system is of component x,

F_x number of failures per year of component x in a
                   year,

F_t sum of the number of failures per year of all
                   the components included by the index x,

R_x repair time (hr/repair) of component x,

RM_z time taken by restoration (hr/res) mode z,

RT_{xy} time taken to restore load point y after its
                   interruption due to the failure of component x,

LPI_y number of interruptions of load point y in a
                   year,

$LPOD_y$ outage duration (hr/repair) per interruption of
                   load point y,

LPAU$_y$ annual unavailability (hr/yr) of load point y,

LPAUE$_y$ contribution to the annual unavailability of
              load point y due to a failure of one of the
              components which when failed results in an in-
              terruption at load point y,

SAIFI   service average interruption frequency index as
              a random variable,

SAIFIE   contribution to SAIFI due to a failure of one
              of the components which when failed results in
              an interruption of at least one load point in
              the system.

SAIDI, SAIDIE, ASAI, ASAIE, ASUI, ASUIE, ENS, ENSE, AENS,
AENSE can be defined in a similar way.

ASSUMPTIONS

The technique is based on the following four assumptions:

(a) the number of failures of a component in a year fol-
    lows a Poisson distribution [2],

(b) the time spent on the repair of a component is very
    small compared to its total operating time,

(c) if the system is in a failed state due to the failure
    of a component, none of its other components can
    fail,

(d) the failure of components are independent of each
    other.

NUMBER OF LOAD POINT INTERRUPTIONS

The number of load point interruptions in a year for
load point y, LPI$_y$, is given by the following expression:

$$LPI_y = \sum_{xy \in x} F_{xy} \qquad\qquad (A4.1)$$

where F_{xy} is a random variable denoting the number of
failures in a year of component xy. The quantity on the
right hand side of Equation (A4.1) is the sum of indepen-

dent Poisson distributed random variables as the component failures are independent of each other (fourth assumption).   It can be proved by random variable theory
that

   "If X_1, X_2, X_3, ..., X_n are independent Poisson random variables with pf's $f(x:\lambda_i)$, (i = 1, 2, ..., n)
   respectively, then the random variable $y = \sum_{i=1}^{n} X_i$, also
   has a Poisson distribution with pf $f(y: \sum_{1}^{n} \lambda_i)$.".

The random variable LPI_y is, therefore, Poisson distributed with an average value equal to the sum of the
failure rates (f/yr) of all the components included by
index xy.

STEP 1 - MOMENTS OF COMPONENT PERFORMANCE PARAMETERS

The random variable F_x, the number of failures of a
component x in a year, follows a Poisson distribution.
The average value of F_x is equal to the failure rate of
the component.   The moments of F_x can, therefore, be determined.   The sum of the failures in a year of all the
components included by the index x, F_t, is also Poisson
distributed. As previously noted the sum of independent
Poisson random variables is also Poisson distributed.
The average value of F_t is equal to the sum of the failure rates of all the components included by x.   The raw
moments of F_t can therefore be obtained.
The probability distributions followed by component
repair times and the system restoration times, are specified as input parameters of the analysis.   The random
variable, RT_{xy} is the time taken to restore the load
point y after its interruption due to the failure of component x, R_x, or one of the restoration times, depending
on the configuration.   Since the moments of R_x and all
the restoration times are known, the moments of RT_{xy} can

be determined.

STEP 2 - EVALUATION OF MOMENTS

Basic Concepts

This section describes the development of techniques for the evaluation of the moments of the reliability indices [1]. The approach used to develop these techniques for any of the reliability indices, y, which in most cases is an intricate function of a number of random variables X_1, X_2, X_3, ... (Equation A4.1) is to disaggregate this function into some basic functions which are easier to handle. The moments of the reliability index y are then obtained by coupling the techniques to evaluate moments of the basic functions. In the analysis of some of the basic functions, it is convenient to deal with raw moments while in others with central moments. The techniques for the basic functions when coupled therefore, may require conversion of raw moments to central moments and vice-versa. The method for evaluating the moments of the basic functions and the formulae to convert the moments from one form to another are illustrated before describing the techniques to evaluate the moments of the reliability indices.

There are three basic functions of random variables required in the analysis. These three functions are, combination of random variables, algebraic function of random variables and the random sum of random variables. It should be noted that the random variables forming these functions have been assumed to be independent.

Combination Of Random Variables

Consider Z to be a function of the random variables Y_1, Y_2, Y_3, ... such that the value taken by Z is equal to the value of any of these random variables and the probability associated with Z taking the value of Y_n is

p_n. Using the conditional probability approach [1]:

$$E[Z^r] = \sum_{i=1}^{n} p_i E[Y_i^r] \qquad (A4.2)$$

The first four raw moments of Z, can therefore be obtained knowing these moments for Y_1, Y_2, Y_3, ... and the associated probabilities p_1, p_2, p_3, ... It should be noted that the sum of these probabilities will be equal to one because any value taken by Z has to be of one of these random variables.

Algebraic Functions Of Random Variables

Let Z be an algebraic function of the random variables X_1, X_2, X_3, ..., X_n, e.g.:

$$Z = h(X_1, X_2, X_3, \ldots, X_n) \qquad (A4.3)$$

If X_i are uncorrelated random variables, it can be proved [4] that:

$$E[Z] = h(E[X_1], E[X_2], \ldots, E[X_n])$$

$$+ \frac{1}{2} \sum_{i=1}^{n} \frac{\delta^2 h}{\delta X_i^2} \mu_2[X_i] \qquad (A4.4)$$

$$\mu_2[Z] = \sum_{i=1}^{n} \left(\frac{\delta h}{\delta X_i}\right)^2 \mu_2[X_i] + \sum_{i=1}^{n} \left(\frac{\delta h}{\delta X_i}\right)\left(\frac{\delta^2 h}{\delta X_i^2}\right) \mu_3[X_i] \qquad (A4.5)$$

$$\mu_3[Z] = \sum_{i=1}^{n} \left(\frac{\delta h}{\delta X_i}\right)^3 \mu_3[X_i] \qquad (A4.6)$$

$$\mu_4[Z] = \sum_{i=1}^{n} \left(\frac{\delta h}{\delta X_i}\right)^4 \mu_4[X_i] + 6 \sum_{\substack{i \ j \\ i<j}} \left(\frac{\delta h}{\delta X_i}\right)^2 \left(\frac{\delta h}{\delta X_j}\right)^2$$

$$\mu_2[X_i]\mu_2[X_j] \qquad (A4.7)$$

All derivatives in these expressions are evaluated at the mean value of the random variables. The expressions

give approximate values of moments since they have been
derived using Taylor series expansion of function h about
the point at which each of the component random variables
take their mean values.   When h is a linear algebraic
function:

$$Z = \sum_{i=1}^{n} a_n X_n \qquad\qquad (A4.8)$$

The values of the derivatives in Equations A4.5 –
A4.8 for this case are:

$$\frac{\delta h}{\delta X_i} = a_i, \quad \frac{\delta^2 h}{\delta X_i^2} = 0 \qquad\qquad (A4.9)$$

Therefore:

$$E[Z] = \sum_{i=1}^{n} a_i E[X_i] \qquad\qquad (A4.10)$$

$$\mu_2[Z] = \sum_{i=1}^{n} a_i^2 \mu_2[X_i] \qquad\qquad (A4.11)$$

$$\mu_3[Z] = \sum_{i=1}^{n} a_i^3 \mu_3[X_i] \qquad\qquad (A4.12)$$

$$\mu_4[Z] = \sum_{i=1}^{n} a_i^4 \mu_4[X_i] + 6 \sum_{\substack{i \ j \\ i<j}} a_i^2 a_j^2 \mu_2[X_i] \mu_2[X_j] \qquad (A4.13)$$

The mean value and second, third and fourth central
moments of a linear algebraic function of random variables can therefore be obtained by knowing the corresponding moments for the random variables which comprise
the function.   The moments obtained in this case, will be
exact as the Taylor series expansion will exactly represent the linear algebraic function as the second and
higher order derivatives are zero.   The variables X_i in
Equations A4.10–13 are uncorrelated.   These equations are

valid for the independent random variables because if the
variables are independent, they are also uncorrelated.

Random Sum Of A Random Variable

Consider a random variable S, such that:

$$S = \sum_{k=0}^{N} X_k \qquad\qquad (A4.14)$$

where N is a discrete random variable,

X_1, X_2, ..., X_k are random variables independent of
N,

and   S = 0 when N takes the value 0.

The random variable S of this type is defined as the
random sum. If random variables X_k are independent and
identically distributed with the first four raw moments
denoted by $E[X]$, $E[X^2]$, $E[X^3]$, $E[X^4]$ and random variable
N with raw moments given by $E[N]$, $E[N^2]$, $E[N^3]$, $E[N^4]$,
then it can be proved that:

$$E[S] = E[X]E[N] \qquad\qquad (A4.15)$$

$$E[S^2] = \{E[X^2] - (E[X])^2\}E[N] + (E[X])^2E[N^2] \quad (A4.16)$$

$$E[S^3] = \{E[X^3] - 3E[X^2]E[X] + 2(E[X])^3\}E[N]$$
$$+ \{3E[X^2]E[X] - 3(E[X])^3\}E[N^2]$$
$$+ \{(E[X])^3\}E[N^3] \quad \text{and} \qquad (A4.17)$$

$$E[S^4] = \{E[X^4] - 4E[X^3]E[X] - 3(E[X^2])^2$$
$$+ 12E[X^2](E[X])^2 - 6(E[X])^4\}E[N]$$
$$+ \{4E[X^3]E[X] + 3(E[X^2])^2 - 18E[X^2](E[X])^2$$
$$+ 11(E[X])^4\}E[N^2] + \{6E[X^2](E[X])^2$$
$$- 6(E[X])^4\}E[N^3] + \{(E[X])^4\}E[N^4] \qquad (A4.18)$$

The first four raw moments of S can be evaluated
knowing these moments for the identically distributed
random variables X_k and the discrete random variable N.

Consider a random variable K, such that the value taken by it is equal to the sum of N independent observations of a random variable X, where N is a discrete random variable. Such a random variable is equivalent to the random variable S, for which the moments are given by Equations A4.15-18. The moments of K can, therefore, be evaluated using the same equations by utilizing the raw moments of X instead of X_i.

Conversion Of Moments

The following formulae can be used to convert central moments to raw moments and vice-versa.

$$\mu_2[X] = E[X^2] - (E[X])^2 \tag{A4.19}$$

$$\mu_3[X] = E[X^3] - 3E[X^2]E[X] + 2(E[X])^3 \tag{A4.20}$$

$$\mu_4[X] = E[X^4] - 4E[X^3]E[X] + 6E[X^2](E[X])^2$$
$$- 3(E[X])^4 \tag{A4.21}$$

$$E[X^2] = \mu_2[X] + (E[X])^2 \tag{A4.22}$$

$$E[X^3] = \mu_3[X] + 3E[X]\mu_2[X] + (E[X])^3 \tag{A4.23}$$

$$E[X^4] = \mu_4[X] + 4E[X]\mu_3[X] + 6(E[X])^2\mu_2[X]$$
$$+ (E[X])^4 \tag{A4.24}$$

Moments Of Reliability Indices

The analytical techniques required to determine the mean and second, third and fourth central moments of the reliability indices are given in this section. The values of the first four moments of the component performance parameters and system restoration times are required in order to determine the moments of the reliability indices using these techniques. In the following analysis the expressions for the first four raw moments of the indices have been developed. The second, third and fourth central moments are obtained utilizing these raw moments and Equations A4.19-21. The probability of a failure event of a given component x in the system, P_x, is given by the ratio of the failure rate of component x and the sum of failure rates of all the components included by index x.

Load Point Outage Duration. The outage duration of any load point y, $LPOD_y$, is defined as the time to restore the supply to the load point y if it is interrupted due to a component failure. The time to restore if load point y has been interrupted due to the failure component x, is RT_{xy}. The load point outage duration is a combination of random variables RT_{xy}. Using Equation A4.2, the expression for the raw moments of $LPOD_y$ is as follows:

$$E[LPOD_y^r] = \sum_{xy\epsilon x} P_{xy}E[RT_{xy}^r] \qquad (A4.25)$$

The probability that the given outage of a load point y is due to the component xy, P_{xy}, can be evaluated as a ratio of the failure rate of component xy and the sum of failure rates of all the components included by index xy.

Load Point Annual Unavailability. The annual unavailability $LPAU_y$ of a load point y, is defined as the total time in a year for which the load point remains out of service. The number of interruptions of the load point y in a year are given by LPI_y. Each of these interruptions are of duration $LPAUE_y$ or $LPOD_y$. Therefore:

$$LPAU_y = \sum_{k=0}^{LPI_y} LPOD_{y_k} \qquad (A4.26)$$

e.g. $LPAU_y$ is a random sum of a random variable $LPOD_y$. The random variable LPI_y follows a Poisson distribution as noted previously. The raw moments of $LPAU_y$ can, therefore, be obtained utilizing these moments of $LPOD_y$ and LPI_y in Equations A4.15-18.

System Average Interruption Frequency Index. SAIFI is defined as the number of customer interruptions per system customer in a year. The index yx includes all those load points which are interrupted if the component x fails. Therefore, the total number of customers interrupted as a result of failure of component x is $\sum\limits_{yx\epsilon y} c_{yx}$. If the total number of customers in the system is represented by c_t, then the number of customer interruptions per customer due to failure of component x is $\sum\limits_{yx\epsilon y} (c_{yx}/c_t)$.

A given component failure event can be due to any of the components included in x. The contribution to SAIFI by such an event is a combination of quantities given by $\sum\limits_{yx\epsilon y} (c_{yx}/c_t)$. The raw moments of:

$$E[SAIFIE^r] = \sum_x P_x[\{ \sum_{yx\epsilon y} (\frac{c_{yx}}{c_t})\}^r] \qquad (A4.27)$$

The raw moments of a constant quantity n is given by:

$$E[n^r] = n^r \qquad (A4.28)$$

The quantity $\sum\limits_{yx\epsilon y} (c_{yx}/c_t)$ is constant. The raw moments of SAIFIE can therefore, be obtained using the raw moments of $\sum\limits_{yx\epsilon y} (c_{yx}/c_t)$ in Equations A4.27. There are F_t number of events of component failures in a year. Each of these events contribute SAIFIE to the value of SAIFI. Therefore SAIFI is a random sum of SAIFIE and is given by:

$$SAIFI = \sum_{k=0}^{F_t} SAIFIE_k \qquad (A4.29)$$

Knowing the raw moments of SAIFIE and of F_t, the raw moments of SAIFI are determined using Equations A4.15-18.

System Average Interruption Duration Index. SAIDI is defined as the number of customer hours lost per system

customer in a year. The contribution to SAIDI by a failure event of a given component x is:

$$(\sum_{yx0\epsilon yx} \frac{c_{yx0}}{c_t})R_x + \sum_z (\sum_{yxz\epsilon yx} \frac{c_{yxz}}{c_t})RM_z$$

It should be noted that the set yxεx includes those
load points which experience an outage due to the failure
of component x. The load points in set yx0εyx are restored after the repair of component x and therefore the
interruption is of duration R_x. The load point in set
yxzεyx are restored by the restoration mode z so their
interruption is of duration RM_z. The possible modes of
restoration are included in set Z. The contribution to
SAIDI by a failure event of the component x given above
is a linear algebraic function of random variables and
its moments are evaluated using Equations A4.11-14. The
moments so obtained will be the mean and the second,
third and fourth central moments. The central moments are
converted to the corresponding raw moments using Equations A4.22-24. The failure event can be of any component included by index x and SAIDIE is therefore a combination of these quantities. Using Equation A4.2:

$$E[SAIDIE^r] = \sum_x P_x E[\{ (\sum_{yx0\epsilon yx} \frac{c_{yx0}}{c_t})R_x$$

$$+ \sum_z (\sum_{yxz\epsilon yx} \frac{c_{yxz}}{c_t})RM_z \}^r] \qquad (A4.30)$$

There are F_t number of component failure events in a
year and each of these events contributes SAIDIE to the
value of SAIDI, therefore:

$$SAIDI = \sum_{k=0}^{F_t} SAIDIE_k \qquad (A4.31)$$

The raw moments of SAIDI are evaluated using Equations A4.15-18.

Average Service Unavailability Index. ASUI is the customer hours lost in the system per customer hour demanded in a year. The contribution to ASUI by a failure event of a given component x is:

$$(\sum_{yx0\epsilon yx} \frac{c_{yx0}}{8760.c_t}) R_x + \sum_z (\sum_{yxz\epsilon yx} \frac{c_{yxz}}{8760.c_t}) RM_z$$

This quantity has terms similar to the corresponding quantity for SAIDI, except for the constant 8760, which denotes the total number of hours in a year. The raw moments of ASUIE are given by the following expression:

$$E[ASUIE^r] = \sum_x P_x E[\{(\sum_{yx0\epsilon yx} \frac{c_{yx0}}{8760.c_t}) R_x$$

$$+ \sum_z (\sum_{yxz\epsilon yx} \frac{c_{yxz}}{8760.c_t}) RM_z\}^r] \qquad (A4.32)$$

There are in total F_t number of component failure events which occur in a year and each contributes ASUIE to ASUI. Therefore:

$$ASUI = \sum_{k=0}^{F_t} ASUIE_k \qquad (A4.33)$$

The raw moments of ASUI are determined using Equations A4.15-18.

Average Service Availability Index. ASAI is the number of customer hours made available per customer hour demanded in the system. The value of ASAI can be specified in terms of ASUI as:

$$ASAI = 1 - ASUI \qquad (A4.34)$$

Therefore:

$$E[ASAI] = E[1 - ASUI] = 1 - E[ASUI] \qquad (A4.35)$$

$$E[ASAI^2] = E[(1 - ASUI)^2]$$

$$= 1 + E[ASUI^2] - 2E[ASUI] \qquad (A4.36)$$

$$E[ASAI^3] = E[(1 - ASUI)^3 = 1 - 3E[ASUI]$$

$$+ 3E[ASUI^2] - E[ASUI^3] \quad \text{and} \qquad (A4.37)$$

$$E[ASAI^4] = E[(1 - ASUI)^4] = 1 - 4E[ASUI]$$

$$+ 6E[ASUI^2] - 4E[ASUI^3] + E[ASUI^4] \qquad (A4.38)$$

The moments of ASAI are obtained using these equations as the raw moments of ASUI are known.

Energy Not Supplied. ENS is defined as total energy (kWhr) lost in the system in a year. The contribution to ENS by a failure event of a given component x is:

$$(\sum_{yx0 \varepsilon yx} l_{yx0}) R_x + \Sigma (\sum_{z \ yxz \varepsilon yx} l_{yxz}) RM_z$$

The term l_{yx0} denotes the load (kW) connected at load point yx0. Using Equations A4.11-14 for the moments of a combination of random variables:

$$E[ENSE^r] = \sum_x P_x \, E[\{(\sum_{yx0 \varepsilon yx} l_{yx0}) R_x$$

$$+ \Sigma (\sum_{z \ yxz \varepsilon yx} l_{yxz}) RM_z \}^r] \qquad (A4.39)$$

Since F_t number of component failure events occur in the system and each of these failures contributes ENSE to ENS:

$$ENS = \sum_{k=0}^{F_t} ENSE_k \qquad (A4.40)$$

The raw moments of ENS are evaluated using Equations A4.15-18.

Average Energy Not Supplied Index. AENS is defined as total energy (kWhr) lost in the system per customer in a year. The expressions for determining the moments of AENSE are as follows. These have been developed in a similar manner to the expressions for the moments of SAIDIE:

$$
E[AENSE^r] = \sum_x P_x E[\{ (\sum_{yx0\epsilon yx} \frac{l_{yx0}}{c_t}) R_x
$$

$$
+ \sum_z (\sum_{yxz\epsilon yx} \frac{l_{yxz}}{c_t}) RM_z \}^r] \tag{A4.41}
$$

$$
AENS = \sum_{k=0}^{F_t} AENSE_k \tag{A4.42}
$$

The raw moments of AENS are evaluated using Equations A4.15-18.

STEP 3 – EVALUATION OF APPROXIMATE PERCENTILES

A family of probability distributions known as the Pearson family can be generated [3] by solving the following differential equation:

$$
\frac{df(x)}{dx} = \frac{(x + a)f(x)}{b_0 + b_1x + b_2x^2} \tag{A4.43}
$$

The form of solution depends on the values of the constants a, b_0, b_1, and b_2 which are related to the first four moments of the variable x. The solution of Equation A4.43 leads to a large number of distribution families, including normal, gamma, beta, etc. The wide diversity of shapes of the Pearson distribution curves allow one to select the most appropriate distribution for a given random variable, based on the knowledge of its first four moments. The procedure for determining the most suitable Pearson distribution is very involved. For

easier use of the Pearson family of curves, a table of
standardized percentiles have been published [4] based on
the solution of Equation A4.43. These tables can be used
to evaluate the estimated percentiles of a given random
variable if its mean value and the second, third, fourth
central moments are given. The procedure for using the
tables involves the following steps:

(a) calculate the value of parameters $\sqrt{\beta_1}$ and β_2 as fol-
    lows:

$$\sqrt{\beta_1} = \frac{\mu_3[X]}{(\mu_2[X])^{3/2}} \qquad\qquad (A4.44)$$

$$\sqrt{\beta_2} = \frac{\mu_4[X]}{(\mu_2[X])^2} \qquad\qquad (A4.45)$$

(b) find the tabulated standardized percentile Z_α for the
    chosen α using $\sqrt{\beta_1}$ and β_2. Use interpolation wher-
    ever required,

(c) the estimated αth percentile is then calculated as
    αth percentile $= E[X] + Z_\alpha \sqrt{\mu_2}[X]$ \qquad (A4.46)

REFERENCES

1. Billinton, R. and Goel, R., "An Analytical Approach
   To Evaluate Probability Distributions Associated With
   Reliability Indices Of Electric Distribution Systems," IEEE Transactions, PWRD-1, No. 3, July 1986,
   pp. 245-251.
2. Billinton, R. and Allan, R.N., "Reliability Evaluation Of Power Systems," Longman, London (England)/
   Plenum Publishers, New York, 1984.
3. Elderton, W.P., "Frequency Curves And Correlation,"
   Cambridge University Press 1938.
4. Hahn, G.J. and Shapiro, S.S., "Statistical Models In
   Engineering," John Wiley and Sons, New York, 1967.

INDEX

Printed in the United Kingdom
by Lightning Source UK Ltd.
128925UK00005B/43/A